Information Fusion and Data Science

Series Editor

Henry Leung, University of Calgary, Calgary, AB, Canada

This book series provides a forum to systematically summarize recent developments, discoveries and progress on multi-sensor, multi-source/multi-level data and information fusion along with its connection to data-enabled science. Emphasis is also placed on fundamental theories, algorithms and real-world applications of massive data as well as information processing, analysis, fusion and knowledge generation.

The aim of this book series is to provide the most up-to-date research results and tutorial materials on current topics in this growing field as well as to stimulate further research interest by transmitting the knowledge to the next generation of scientists and engineers in the corresponding fields. The target audiences are graduate students, academic scientists as well as researchers in industry and government, related to computational sciences and engineering, complex systems and artificial intelligence. Formats suitable for the series are contributed volumes, monographs and lecture notes.

More information about this series at http://www.springer.com/series/15462

Tania Cerquitelli · Nikolaos Nikolakis ·
Niamh O'Mahony · Enrico Macii ·
Massimo Ippolito · Sotirios Makris

Editors

Predictive Maintenance in Smart Factories

Architectures, Methodologies, and Use-cases

 Springer

Editors
Tania Cerquitelli
Department of Control and Computer
Engineering
Politecnico di Torino
Turin, Italy

Nikolaos Nikolakis
Laboratory for Manufacturing Systems
and Automation
University of Patras
Patras, Greece

Niamh O'Mahony
Dell EMC Research Europe
Cork, Ireland

Enrico Macii
Interuniversity Department of Regional
and Urban Studies and Planning
Politecnico di Torino
Turin, Italy

Massimo Ippolito
COMAU S.p.A.
Grugliasco, TO, Italy

Sotirios Makris
Laboratory for Manufacturing Systems
and Automation
University of Patras
Patras, Greece

ISSN 2510-1528 ISSN 2510-1536 (electronic)
Information Fusion and Data Science
ISBN 978-981-16-2942-6 ISBN 978-981-16-2940-2 (eBook)
https://doi.org/10.1007/978-981-16-2940-2

This Springer imprint is published by the registered company Springer Nature Singapore Pte Ltd.
The registered company address is: 152 Beach Road, #21-01/04 Gateway East, Singapore 189721, Singapore

Preface

In the last years, Information Technology services and components are becoming pervasive inside production lines, devices, and control systems, transforming traditional manufacturing and shop floor environments into the fully digital, interconnected factory of the future.

The real-time collection of a growing amount of data from factories is paving the way to the creation of *manufacturing intelligence* platforms. *Predictive maintenance*, that is, the ability to predict equipment critical conditions before their occurrence disrupts the production line, is among the most important expressions of manufacturing intelligence. By leveraging innovative computing paradigms from the Internet of Things to cloud computing and data-driven methodologies, effective predictive diagnostic tools can be built and delivered as-a-service. Ultimately, this enables industrial stakeholders to take control of their assets and design smart and efficient maintenance plans that increase productivity while at the same time reducing costs. All industries, regardless of sector and size, can benefit from such solutions; however, integrating data-driven methodologies in shop floor environments is still a complex task, which significantly hinders their adoption, especially from small and medium manufacturing enterprises.

This book presents the fruitful outcomes of the European project "SERENA," involving fourteen partners, including renowned academic centers, IT companies, and industries at the international level. The project addresses the design and development of a plug-n-play end-to-end cloud architecture, enabling predictive maintenance of industrial equipment to be easily exploited by small and medium manufacturing companies with minimal data-analytics experience.

This book offers design principles for data-driven algorithms, architectures, and tools for enhancing intelligence in industrial environments. The book is divided into two parts to ease readability, tailored to readers of different backgrounds. First Part presents more technical content, while Second Part real industrial use cases.

The integrated solution developed with the SERENA project overcomes some of the most challenging issues associated with predictive maintenance (e.g., architecture configuration and deployment, collection of large-scale datasets with a known failure status, and real-time assessment of predictive model performance) by exploiting self-configuring tools and techniques that make their adoption feasible by a wide range of industries. The solution addresses the complete data lifecycle from data collection to value exploitation. It integrates a wide range of innovative data-driven methodologies that have been designed specifically to streamline the prognostics of IoT industrial components, the characterization of equipment health status and operating conditions to generate early warnings, and the forecasting of future equipment degradation. The data-driven models are not only capable of self-configuring, thus requiring limited data analytics expertise, but once deployed, are capable of continuously self-assessing their performance, raising a warning when their predictions are no longer accurate, and the model may need to be retrained to cope, e.g., with evolving operating conditions. All these components are integrated using a scalable methodology-as-a-service approach on a mixed cloud-edge analytics platform.

All proposed solutions are thoroughly compared and discussed with respect to the state-of-the-art and state-of-the-practice methodologies and architectures. Furthermore, methodological approaches and architectures proposed in the context of SERENA have been thoroughly experimentally evaluated on real data sets collected in a variety of industrial environments, including robotics, white goods, manufacturing of elevator cabins, metrological equipment, and steel production. An in-depth discussion of the experimental results is also included for each proposed use case.

Perspectives and new opportunities to address open issues on predictive maintenance conclude each chapter with some interesting future research activities.

We hope readers will find the book of interest and that the content will inspire researchers and practitioners to develop creative, successful, and technological-advanced research!

Turin, Italy   Tania Cerquitelli
Patras, Greece   Nikolaos Nikolakis
Cork, Ireland   Niamh O'Mahony
Turin, Italy   Enrico Macii
Grugliasco, Italy   Massimo Ippolito
Patras, Greece   Sotirios Makris
April 2021

Acknowledgements The editors would like to thank all people who contributed to the SERENA project and inspired the content included in the book.

This research has been partially funded by the European project "SERENA—VerSatilE plug-and-play platform enabling REmote predictive mainteNAnce" (Grant Agreement: 767561).

Contents

Contributors

Andrea Gómez Acebo TRIMEK, Madrid, Spain

Kosmas Alexopoulos Laboratory for Manufacturing Systems and Automation, University of Patras, Patras, Greece

Salvatore Andolina SynArea Consultants s.r.l., Turin, Italy

Daniele Apiletti Department of Control and Computer Engineering, Politecnico di Torino, Turin, Italy

Xanthi Bampoula Laboratory for Manufacturing Systems and Automation, University of Patras, Patras, Greece

Florian Bartholomauss oculavis GmbH, Aachen, Germany

Andrea Bellagarda Department of Control and Computer Engineering, Politecnico di Torino, Turin, Italy

Paolo Bethaz Department of Control and Computer Engineering, Politecnico di Torino, Turin, Italy

Valeria Boldosova Prima Power, Seinäjoki, Finland; University of Vaasa, Vaasa, Finland

Davide Cannizzaro Department of Control and Computer Engineering, Politecnico di Torino, Turin, Italy

Tania Cerquitelli Department of Control and Computer Engineering, Politecnico di Torino, Turin, Italy

Guido Coppo SynArea Consultants s.r.l., Turin, Italy

Claudia De Vizia Department of Control and Computer Engineering, Politecnico di Torino, Turin, Italy

Santa Di Cataldo Department of Control and Computer Engineering, Politecnico di Torino, Turin, Italy

Pietro Greco Research & Innovation, Engineering Ingegneria Informatica S.p.A., Palermo, Italy

Jani Hietala VTT Technical Research Centre of Finland, Espoo, Finland

Massimo Ippolito COMAU S.p.A., Grugliasco, TO, Italy

Sven Jung Fraunhofer Institute for Production Technology IPT and Laboratory for Machine Tools and Production Engineering (WZL) of RWTH Aachen University, Aachen, Germany

Petri Kaarmila VTT Technical Research Centre of Finland, Espoo, Finland

Enrico Macii Interuniversity Department of Regional and Urban Studies and Planning, Politecnico di Torino, Turin, Italy

Sotirios Makris Laboratory for Manufacturing Systems and Automation, University of Patras, Patras, Greece

Angelo Marguglio Research & Innovation, Engineering Ingegneria Informatica S.p.A., Palermo, Italy

Simone Monaco Department of Control and Computer Engineering, Politecnico di Torino, Turin, Italy

Lucrezia Morabito COMAU S.p.A., Grugliasco, TO, Italy

Lia Morra Department of Control and Computer Engineering, Politecnico di Torino, Turin, Italy

Chiara Napione COMAU S.p.A., Grugliasco, TO, Italy

Nikolaos Nikolakis Laboratory for Manufacturing Systems and Automation, University of Patras, Patras, Greece

Matteo Orlando Department of Control and Computer Engineering, Politecnico di Torino, Turin, Italy

Niamh O'Mahony Dell EMC Research Europe, Cork, Ireland

Jari Pakkala Prima Power, Seinäjoki, Finland

Simone Panicucci COMAU S.p.A., Grugliasco, TO, Italy

Stefano Paradiso Stellantis, Turin, Italy

Edoardo Patti Department of Control and Computer Engineering, Politecnico di Torino, Turin, Italy

Pierluigi Petrali Whirlpool EMEA, Biandronno, VA, Italy

Martin Plutz oculavis GmbH, Aachen, Germany

Massimo Poncino Department of Control and Computer Engineering, Politecnico di Torino, Turin, Italy

Eetu Puranen KONE, Hyvinkää, Finland

Olli Saarela VTT Technical Research Centre of Finland, Espoo, Finland

Riku Salokangas VTT Technical Research Centre of Finland, Espoo, Finland

Roberta Sampieri Stellantis, Turin, Italy

Robert H. Schmitt Fraunhofer Institute for Production Technology IPT and Laboratory for Machine Tools and Production Engineering (WZL) of RWTH Aachen University, Aachen, Germany

Georgios Siaterlis Laboratory for Manufacturing Systems & Automation, University of Patras, Patras, Greece

Robert Siegburg Fraunhofer Institute for Production Technology IPT and Laboratory for Machine Tools and Production Engineering (WZL) of RWTH Aachen University, Aachen, Germany

Silvia de la Maza Uriarte TRIMEK, Madrid, Spain

Peter van Wilgen VDL Weweler bv, Apeldoorn, The Netherlands

Antonio Giuseppe Varrella Department of Control and Computer Engineering, Politecnico di Torino, Turin, Italy

Giuseppe Veneziano Research & Innovation, Engineering Ingegneria Informatica S.p.A., Palermo, Italy

Francesco Ventura Department of Control and Computer Engineering, Politecnico di Torino, Turin, Italy

Antonio Ventura-Traveset TRIMEK, Barcelona, Spain

Acronyms

ADC	Analog to Digital Converter
AI	Artificial Intelligence
AIC	Akaike's Information Criterion
AR	Augmented Reality
BA	Blowing Agents
CART	Classification and Regression Tree
CBM	Condition-Based Maintenance
CM	Condition Monitoring
CMM	Coordinate Measurement Machine
CMMS	Computerised Maintenance Management System
CPS	Cyber-Physical System
CRIS	Common Relational Information Model
DH	Decision Horizon
DL	Deep Learning
DS	Descriptor Silhouette
ECS	Elastic Cloud Storage
ERP	Enterprise Resource Planning
FMEA/FMECA	Failure Mode and Effects (or and Criticality) Analysis
GMM	Gaussian Mixture Mode
GPIO	General-Purpose Input Output
GPU	Graphical Processing Unit
HDFS	Hadoop Distributed File System
HMI	Human-Machine Interface
ICT	Information and Communication Technologies
IIoT	Industrial Internet of Things
IoT	Internet of Things
IRI	Internationalised Resource Identifier
ISTEP	Integrated Self-Tuning Engine for Predictive maintenance
JSON	JavaScript Object Notation
JSON-LD	JavaScript Object Notation for Linked Data
KNN	K-Nearest Neighbors
KPI	Key Performance Indicator

LSTM	Long Short-Term Memory
MAAPE	Mean Arctangent Absolute Percentage Error
MES	Manufacturing Execution System
ML	Machine Learning
MNA	Maximum Number of Alternatives
OEM	Original Equipment Manufacturer
OSA-EAI	Open Systems Architecture for Enterprise Application Integration
PdM	Predictive Maintenance
PLC	Programmable Logic Controller
PSBB	Punching-Shearing-Buffering-Bending
PvM	Preventive Maintenance
RCFA	Root Cause Failure Analysis
RMSE	Root Mean Square Error
RNNs	Recurrent Neural Networks
RPCA	Reverse Proxy Certification Authority
RUL	Remaining Useful Life
SERENA	VerSatilE plug-and-play platform enabling REmote predictive mainteNAnce
SME	Small and medium-sized enterprise
SOA	Service-Oriented Architecture
SPI	Serial Peripheral Interface
SR	Sampling Rate
SSL	Secure Sockets Layer
TLS	Transport Layer Security
TPU	Tensor Processing Unit
UI	User Interface
VPN	Virtual Private Network
VR	Virtual Reality
XML	eXtensible Markup Language

Methodologies and Enabling Technologies

Industrial Digitisation and Maintenance: Present and Future

Massimo Ippolito, Nikolaos Nikolakis, Tania Cerquitelli, Niamh O'Mahony, Sotirios Makris, and Enrico Macii

Abstract In recent years various maintenance strategies have been adopted to maintain industrial equipment in an operational condition. Adopted techniques include approaches based on statistics generated by equipment manufacturers, human knowledge, and intuition based on experience among others. However, techniques like those mentioned above often address only a limited set of the potential root causes, leading to unexpected breakdown or failure. As a consequence, maintenance costs were considered a financial burden that each company had to sustain. Nevertheless, as technology advances, user experience and intuition are enhanced by artificial intelligence approaches, transforming maintenance costs into a company's strategic asset. In particular, for manufacturing industries, a large volume of data is generated on a shop floor as digitisation advances. Combining information and communication technologies (ICT) with artificial intelligence techniques may create insight over production processes, complement or support human knowledge, revealing undetected anomalies and patterns that can help predict maintenance actions. Consequently, the

M. Ippolito
COMAU S.p.A., Grugliasco, TO, Italy
e-mail: massimo.ippolito@comau.com

N. Nikolakis (✉) · S. Makris
Laboratory for Manufacturing Systems and Automation, University of Patras, Patras, Greece
e-mail: nikolakis@lms.mech.upatras.gr

S. Makris
e-mail: makris@lms.mech.upatras.gr

T. Cerquitelli
Department of Control and Computer Engineering, Politecnico di Torino, Turin, Italy
e-mail: tania.cerquitelli@polito.it

N. O'Mahony
Dell EMC Research Europe, Cork, Ireland
e-mail: Niamh.Omahony@dell.com

E. Macii
Interuniversity Department of Regional and Urban Studies and Planning,
Politecnico di Torino, Turin, Italy
e-mail: enrico.macii@polito.it

© Springer Nature Singapore Pte Ltd. 2021
T. Cerquitelli et al. (eds.), *Predictive Maintenance in Smart Factories*,
Information Fusion and Data Science,
https://doi.org/10.1007/978-981-16-2940-2_1

3

company yields a reduction of unexpected breakdowns, production stoppages, and production costs. The outcomes are significant but selecting an appropriate data-driven method that can generate helpful and trustworthy results is challenging. It is mainly affected by the quality of the available data and the capability to understand the process under analysis correctly. This chapter reviews architectures for data management and data-driven methodologies for enabling predictive maintenance policies. Then follows the presentation of integrated solutions for predictive analytics to conclude with the main challenges identified and future outlook.

1 Introduction

The advent of the Industry 4.0 brought a revolution in the manufacturing sector, introducing new challenges and opportunities related to Big Data analysis through machine learning techniques. Thanks to sensors installed directly on machinery, companies are now able to continually monitor production activity, collecting a massive amount of data that describe the machinery's behavior through time. The effective processing and analysis of the collected data can provide accurate and valuable results that reveal underlying patterns, identify anomalies not easily detected by a human eye, support the deterioration of the machinery, and support the company's decision-making process about its maintenance activities.

In this context, innovative ICT architectures and data-driven methodologies are needed to effectively and efficiently support predictive diagnostics, allowing industrial stakeholders to plan maintenance operations by using data-driven methodologies-as-a-service efficiently. Although industries need such innovative solutions, the widespread adoption of data-driven methodologies remains challenging in industrial environments. The challenges are further reinforced for small and medium-sized enterprises (SMEs) considering the capital cost required to digitize their processes and benefit from them. Nevertheless, several efforts are underway towards making the use of advanced digital solutions more straightforward and more affordable for SMEs, requiring less capital investment from their end. Nevertheless, the benefits mentioned above are already recognized, and companies adopt similar advanced solutions in their production processes. In fact, one of the twelve "Manufacturing Lighthouses" factories identified by the World Economic Forum is an SME [1], showing how it is rather a matter of paradigm shift than of investments. Stakeholders of the private and public sector recognise the potential benefits of the Industry 4.0 revolution in manufacturing; in many countries national platforms and private-public partnerships have been created to increase awareness, support development of new use-cases and enable collaboration between research institutes and private organisations. Results are encouraging, as 70% of industrial organisations are either piloting Industry 4.0 solutions in manufacturing or using these technologies on a large scale. SMEs, however, can significantly benefit from novel Industry 4.0 technologies by increased flexibility of their production processes under such a smarter production paradigm [2].

The collection of a growing amount of data in factories paved the way to create intelligence over those data and benefit. Predictive maintenance, the ability to identify equipment critical conditions before their actual occurrences, is one type of intelligent practices that has received increasing attention in recent years. Despite the existence of various other maintenance strategies, already used widely in practice, maintenance needs can cause costly disruptions in the manufacturing process. In this context the condition based approach is considered as an effective [3], but also complex to implement and integrate to a production system [4] alternative.

With predictive analytics, however, repair and maintenance activities can be managed more smartly. For example, if needed, they can be prioritised and allocated to pre-planned outages based on real-time probabilities of expected or potential future failures, thus bringing the maintenance costs down. Hence, the predictive maintenance strategy can safely be supported, resulting in savings in both time and costs, due to reduced production downtime [5]. In case this is combined with logistics and maintenance stock parts, then the overall production costs can be reduced even further. Last but not least, predictive maintenance techniques such as vibration and thermal monitoring along with Reliability techniques such as Failure Modes and Effects Analysis (FMEA) and Root Cause Failure Analysis (RCFA) [6, 7] can result in bottom-line savings through early detection and maintenance, preventing failures from disturbing the production process.

However, a purely technology-driven approach to these solutions may easily result in failure; without a clear business objective, the deployment of numerous Internet of Things devices within an industrial plant could lead to an investment without significant return, and equally important, without discovering significant insights from collected data to create business value. The kind of desired data-driven insights tend to determine the appropriate ICT solutions to be adopted by a company. Hence, digitisation and adoption of advanced technologies needs to go hand in hand with a wider transformation of an enterprise's business model and practices.

Considering the emergent need for data-driven analytics and the increased adoption of the as-a-service models, the European project SERENA, funded under the Horizon 2020 framework, introduces a platform for facilitating manufacturers' maintenance needs with a clear focus on supporting predictive maintenance strategies. As part of it, versatile industrial use cases could be analysed considering two discrete needs of the manufacturing sector; equipment providers and equipment consumers or else factories.

After the identification of the industrial needs, we identified data-driven techniques in the context of:

- enabling condition monitoring for providing advanced services to customers, or
- enhancing existing preventive maintenance strategies applied on production equipment or lines with predictive ones.

With respect to enabling the aforementioned techniques with the least disruption of the companies' practices, the as-a-service model was adopted, leading to the design and implementation of a web platform that could be scalable, resilient and deployed in different ways: *at a cloud*, *on-premise, or*, *hybrid*, while facilitating the provision of latest research and development services in the context of condition monitoring and predictive maintenance to inherently heterogeneous industrial sectors. In particular the industrial sectors considered in the context of the SERENA project [8] include the following manufacturing companies: (1) Robotic equipment manufacturer, (2) Metrology equipment manufacturer, (3) Elevators producer, (4) Steel parts producer, (5) White goods manufacturing company.

To this end and to address the requirements for remote condition-monitoring of industrial assets, the following functionalities have been identified as key enablers of facilitating a transition to predictive maintenance approaches in the industry:

- Smart acquisition mechanisms at the edge for collecting data from heterogeneous assets (e.g., robots, machines, welding guns, PLCs, external sensors)
- Increased connectivity for remote data management of the huge data volumes generated by shopfloor data sources
- A software platform supporting predictive maintenance as-a-service as a result of data-driven methodologies,
- Technology-neutral middleware with security capabilities
- Versatile deployments, addressing different needs
- Scalability of functionalities and data processing capabilities

The approaches mentioned above enabled the creation of the SERENA system where each item mentioned above has been implemented into software services and seamlessly integrated into its leading platform.

This chapter is organised as follows: Sect. 2 presents a literature review in the context of intelligent manufacturing with a specific focus on data management architectures, data-driven methodologies for analytics and integrated solutions. Next the key challenges identified are presented in Sect. 3, to conclude with Sect. 4 presenting the trends in the context of predictive maintenance along with the vision of the SERENA project.

2   Literature Review

In the last decade a rich landscape of research and development activities has been carried out in the context of Industry 4.0. Research works can be classified into four main complementary categories:

- Maintenance approaches, discussed in Sect. 2.1.
- Data management architectures, discussed in Sect. 2.2.
- Data analytics methodologies and algorithms are presented in Sect. 2.3, tackling (i) anomaly detection, (ii) predictive analytics, and (iii) Remaining Useful Life (RUL) estimation.
- Integrated solutions are described in Sect. 2.4.

2.1 Maintenance Approaches

Maintenance activities have been traditionally reactive, with corrective operations performed after a failure was observed. This, however, caused significant losses for the production system [9]. As a response, Preventive Maintenance (PvM) approaches were established [10]. PvM approaches objective is to proactively perform fixed in time routine maintenance operations or inspections in order to prevent breakdowns, considering the importance of an uninterrupted production process [11, 12]. In this context, frequent causes of failures are identified in advance, either from experiments or out of experience. Nevertheless, the modelling of a machine's degradation is a complex process [13].

Thus, Condition Monitoring (CM) approaches were introduced providing huge amounts of data of the actual operational status of a machine, known as Big Data [14–16]. However, the identification of the correct set of parameters which are related to the operational status of the machine and its degradation process is not an easy case, especially when considering the number and heterogeneity of equipment on a shopfloor [17–19].

For this reason, Predictive Maintenance (PdM) approaches were introduced allowing the examination of these large-scale data sets [20]. PdM approaches, find out correlations and patterns in order to predict and prevent conditions that can be harmful to the operational lifetime of the production equipment. The ultimate ambition though, is the prolongation of the maintenance procedure and simultaneously the estimation of Remaining Useful Life (RUL) of the industrial equipment [21]. RUL estimation can be achieved through statistical methods and probabilistic models that apply to the available data without depending on physics of the underlying degradation process [22]. The main difference can be set upon the targeted outcome. For statistical learning, the aim is understanding the data correlation and inferences, while in Machine Learning (ML) the outcome of the evaluation is important without requiring a clear understanding of the underlying data and their interaction [23]. Purpose of statistical learning can be the mathematical analysis of the data values in order to discover relations among them and draw inferences to approximate the reality, while, ML approaches are based upon statistical methods in order to teach a machine approximate to real conditions [24, 25].

However, the performance of PdM applications depends on the appropriate choice of the ML method, as various techniques have been investigated until now [26, 27]. Bayesian networks can be used for diagnosing and predicting faults in large manufacturing dataset with little information on the variables, presenting computational learning and speed issues [28]. On the other hand, Convolutional Neural Networks present an excellent performance and a low computational cost [29]. However, Recurrent Neural Networks (RNNs) contain feedback loops and are more suitable for sequential data such as time-series data [30–32]. Nevertheless, since RNNs present some issues on long-term RUL predictions, Long Short-Term Memory (LSTM) networks are preferred [33], presenting, however lack of efficiency because of sensitivity to dataset changes [34]. As a consequence, hybrid models combine more than one network in order to overcome these kind of defects [35]. LSTMs-Autoencoder networks, for example, are presented in [36] as a deep learning approach for RUL estimation of a hot rolling milling machine. Finally, Transformer-based approaches have recently received increased attention for forecasting time series data as they seem to outperform the other ML models [37, 38].

2.2 Data Management Architectures

An emerging challenge of modern industries is to effectively collect, process and analyse large amounts of data, even in real time. To this aim, various data management architectures have been proposed in years [39–46] based on Big Data frameworks. These approaches may differ in some aspects and may complement each other or even demonstrate similarities in their methods. In turn this may address the need for efficient data storage, effective communication, and knowledge extraction mechanisms tailored to the individual characteristics of each industrial case. In particular, the work in [39] presents how recent trends in Industry 4.0 solutions influence the development of manufacturing execution systems, while in [40] authors present a framework facilitating scalable and flexible data analytics towards realising real-time supervision systems for manufacturing environments.

Early software solutions for production systems were mainly based on monolithic architectures providing low flexibility, adaptability, and scalability, thus increasing the complexity to upgrade as well as the maintenance costs. A step towards more flexible architectures was based on Service-Oriented Architectures (SOA), overcoming some previous drawbacks, such as maintainability and flexibility. Unfortunately, these systems could not guarantee high levels of flexibility and modularity required by modern business models. With the advent of cyber-physical production systems, new business models focusing on the as-a-service strategy emerged. As a consequence, versatile strategies based on micro-services architectures were proposed [47]. As discussed in [48], monolithic applications can be decomposed into micro-services to implement lightweight and scalable strategies for achieving better management of services distribution and operation. Nevertheless, an emerging concern is related to the data stewardship and ownership, especially as data are slowly getting linked to value. Data management architectures, at the beginning, were mostly based on cloud resources to enable on-demand services and scalability to future needs. However, this structure gave the cloud provider complete authority over the data. One step towards more privacy-preserving data was represented by fog computing engines [49], enabling on-demand computation to selectively exploit edge devices or cloud resources based on the kinds of jobs to be executed. A review of the recent fog computing trends and applications is provided in [50], where among other challenges the application service, resources management, and communication among layers while enabling security and privacy features, are highlighted.

A parallel research effort was devoted to the main challenges associated with effectively integrating existing production software, legacy systems and advanced

Internet of Things (IoT) technologies [41]. To this aim, the enabling technologies are based on the virtualization strategy and a cloud-based micro-service architecture. The latter technology is mainly exploited in different research activities to deal with big data applications [42, 43].

A step towards more heterogeneous architectures is discussed in [51] where virtualisation technologies combined with IEC61499 function blocks enabled the holistic orchestration of a production station. As a result, the configuration or re-configuration is highly dependent on software resulting in increased automation levels and flexibility.

Furthermore, the authors in [44] presented a Big Data analytics framework, capable of providing a health monitoring application for an aerospace and aviation industry. The studies in [45, 46] use open source technologies such as Apache Spark and Kafka to implement a scalable architecture capable of processing data both online and offline.

2.3 Data-Driven Analytics: Methodologies and Algorithms

The advent of the Industry 4.0 brought a revolution in the manufacturing sector, introducing new challenges and opportunities related to data analysis as enabled by machine learning, data mining and statistical processing techniques. Advanced sensors installed directly on the machinery, allow the constant monitoring of the production activity, collecting a huge amount of data that describes the behaviour and performance of monitored equipment over time. In addition, the effective processing and analysis of collected data, can facilitate a company's decision making process. Among the most beneficial approaches that the collection and analysis of industrial data has brought in a manufacturing context, there are the *predictive maintenance* strategies. Such approaches allow the analysis of a huge amount of data with different algorithms and addressing different objectives and requirements. Predictive maintenance strategies can be implemented through different data analytics methodologies, including *anomaly detection*, *predictive analytics*, and *RUL*, discussed in Sects. 2.3.2, 2.3.2, and 2.3.3, respectively.

To guarantee the maintenance process chain, the assessment of the current equipment condition is an important issue to be addressed. To this end, predictive maintenance strategies focus on identifying possible malfunctions over time and estimating the *RUL*, for assessing the operational life time of a machine.

2.3.1 Anomaly Detection

Different anomaly detection strategies have been proposed based on a joint exploitation of raw data and smart trend filtering [52]. The main drawback of the existing strategies is the need of raw data of proper quality that include both normal and abnormal working conditions. Unfortunately, in many real-life settings such data are unavailable. Methods, to detect noise and outliers, for data quality improvements are discussed by [53].

2.3.2   Predictive Analytics

The topic of predictive maintenance in a big data environment is also addressed in [54], where, with the purpose of monitoring the operation of wind turbines, a data-driven solution deployed in the cloud for predictive model generation is presented.

The spread of predictive maintenance in recent years is wide. In [55] a systematic review of the literature is performed to analyze academic articles about predictive maintenance and machine learning in an Industry 4.0 context, from 2015 to 2020. In these scenarios, historical data set has a fundamental role in order to obtain real time and satisfactory results even on new data. For this reason, steps such as features engineering and data labelling (if the label is not already present) are particularly important and can greatly influence the quality and accuracy of results obtained, as well as the selection of the prediction model.

2.3.3   Estimation of the Remaining Useful Life

The estimation of a component's RUL can be defined as the period from a given starting time up to the point it deteriorates to a level that is no longer operating within the limits evaluated by its producer and for a specific process. This evaluation as well as a preventive maintenance plan, are usually result of extensive tests conducted by the Original Equipment Manufacturer (OEM). These tests, even though covering a wide area of parameters and values to provide the best possible estimations, cannot possibly address all possible use cases and degradation factors that may impact the behaviour of a machine. Thus, it is not that following the preventive maintenance plan provided by an OEM leads to over-maintenance, which is associated with increased cost. On the other hand, if the maintenance deviates a lot from that preventive maintenance plan, it may be the cause of unexpected breakdowns and thus increased production costs. Hence, the estimation of the degradation of machine under its operating environment becomes a factor that could reduce costs, given a certain level of accuracy in its estimation. Usually, the RUL value is adopted to represent and quantify the estimation of a machine's degradation.

In this context, several methodologies to estimate the status of machine's degradation, in terms of a machine's RUL, have been devised [56]. State-of-the-art methods [57–59] are mainly based on complex statistical analysis to map a value with physical meaning for the maintenance personnel, responsible for further actions.

An adaptive skew-Wiener process model is discussed in [60] and validated in a use case concerning the ball bearings in rotating electrical machines. In the same study, the need to incorporate stochastic phenomena is highlighted. In this context, a review on Wiener-process-based methods for degradation data analysis, modelling and RUL estimation is provided in [61].

Moreover, a state space model for prognostics using a mathematically derived Kalman filtering approach is discussed in [62]. The proposed model is shown to result in lower RUL estimation errors in comparison to other approaches defined in the same study and based on a use case concerning on a battery RUL prediction.

Nevertheless, mathematical modelling is a complex process. However, with the rapid development of information and communication technology, data-driven approaches are gaining momentum, promising increased accuracy in RUL estimation. An approach for estimating the RUL of rolling element bearings is discussed in [63]. Relevance vector machines were used with different kernel parameters while experimental results support the effectiveness of the proposed hybrid approach.

Considering the enormous amount of data generated in manufacturing shop floors, Artificial Intelligence (AI) and in particular deep learning approaches have been widely investigated with promising results [64, 65]. Deep recurrent neural networks are discussed in [66, 67]. In both studies the need for a large dataset and sufficient training data is highlighted. LSTM recurrent neural network is another widely used AI method with reports validating its performance [68, 69].

In addition, novel approaches based on data-driven techniques are investigated, such as transformer-based neural networks [70] and digital twin models [71].

2.4 *Integrated Solutions*

A parallel research effort has been devoted to proposing integrated solutions providing both an architectural engine and data-driven methodologies [43, 46]. The work in [46] proposed a distributed engine, enabling predictive maintenance based on a dynamic combination of native interpretable algorithms to provide human-readable knowledge to the end user.

However, integrated solutions have been customized either to a specific use case or to a given data analytics task, such as predictive analytics driven by interpretable models [46], forecasting activities in the smart-city context [43]. More general solutions should be devised with the final aim to easily integrate them in existing industrial solutions.

3 Key Challenges Identified

The decreasing cost of electronics in tandem with their ever-increasing capabilities made sophisticated embedded electronics cost-effective for various applications. As a result, complete historical trends can be stored in digital means and processed by advanced techniques, revealing underlying patterns and detecting anomalies. Thus, data-driven techniques pave the way for novel strategies bringing additional benefits for adopters. In the case of manufacturers, this can bring down production costs via multiple approaches, predictive analytics, and the resulting adoption of predictive maintenance strategies. Nevertheless, while the research community has provided a broad spectrum of advanced methodologies for data analysis and prediction, there is no straightforward way to select the most appropriate. Consequently, extensive experimental validation is needed to compare various techniques and evaluate the impact of multiple parameters that may affect the overall algorithm's performance.

The availability of accurate data is of paramount importance. A proper dataset should include all possible features modeling the machine behavior to be used as training data. An adequate portion of potential defects and failure data would be available to predict future failures correctly. Unfortunately, such data are not available in most cases. Many companies do not keep detailed records of their malfunctions or sensors integrated into their machinery do not monitor some phenomena. Consequently, the training of a predictive model with a proper high-quality dataset is exceptionally challenging. As a result, the trained models' performance cannot achieve their theoretical potential in terms of accuracy of prediction and real-world deployment. In this context, real-life settings are also concerned with the aspects of security and data sovereignty. The importance of these factors is only reinforced as data are gradually gaining value in today's digital transformation.

Furthermore, adopting a predictive model for potential failures' estimation will provide a probability, resulting in a time interval between the estimated failure and the actual functional failure. It is desired from a cost perspective to minimize that interval, but the proper breakdown is only estimated. Thus, unless a certain level of reliability is established upon the predictive model's accuracy, conservative approaches will still be followed, assuming the interval mentioned above being lower than the actual one or resulting in more frequent inspections and costs.

While predictive analytics with real-time condition monitoring can depend on robust infrastructure and achieve predictions of high accuracy and reliability for a well-trained model, this still requires significant capital investment on infrastructure and human experts. Such investments are not possible for many small and medium enterprises and even for large enterprises.

Finally, the insight created by predictive analytics, regardless of the achieved accuracy, is of little value if not combined with the proper mitigation actions and an appropriate communication channel to the maintenance personnel.

4 Outlook

As discussed in this chapter, in the context of smart manufacturing, one of the most important needs that should require further investigation and effective integration of cutting-edge solution is to control remotely the factory equipment status applying predictive maintenance techniques, by automatically having answers on specific questions—Where is the problem? In which station, in which machine and/or in which component?—enabling prevention of failures to happen and allowing improvements in estimating the RUL of a component under given or planned conditions.

The current state-of-the-art manufacturing context shows a rich landscape of data-driven solutions to help industries become smart. However, different interesting challenges are still open, and they represent opportunities to be addressed by the research and industrial community in the short future. From the industrial point of view, one of the most critical challenges is to provide the opportunity/means to use data-driven methodologies and architecture without high-level expertise. Hence, the following can be supported: (i) **architectures should be released as deployable and independent components to be self-configurable and easily integrated with existing industrial infrastructure**. Furthermore, (ii) **data-analytics methodologies should be released as deployable services to be easily integrated into either existing infrastructure or innovative and self-configurable architecture**. In addition, for some analytics tasks that are usually very custom, i.e., data cleaning, the deployable units should propose a generic strategy, easily customised, using a step-by-step technique that supports the end-user in configuring the critical building tasks.

There is still room for improvements in the research area of data analytics tailored to different aspects of manufacturing data as discussed in [72].

Self-evolution of predictive models. Data-driven models' validity is mainly affected by the data drifts on the collected measurements, mainly dependent on machine activities and failures. In the literature, only a few solutions exist to passively or actively update the model. Innovative and self-configured solutions should be devised to automatically assess the model's validity and make decisions on how and when the model must up updated.

Human-readable prediction models. Real-life settings, such as manufacturing context, need interpretable and transparent models to understand the algorithm's decisions easily. For complex prediction tasks on media data, e.g., images, videos, or sounds captured on Industry 4.0 scenarios, black-box algorithms such as neural networks are needed to reach high accuracy. Unfortunately, their generated models are unreadable since algorithms' internal mechanisms are hidden, i.e., the main relationships between the input data and the algorithm's outcomes are unknown. Innovative methodologies that should run on top of black-box models are needed to streamline the exploitation of Industry 4.0 applications' neural-network model.

Class imbalance. Manufacturing data are usually characterized by many samples related to ordinary behavior of the machine/line under monitoring and a very few samples describing failure events. Although this condition is highly desired, it is not an excellent condition for data analytics techniques. The latter infers the main hidden and latent relationships among data from a huge set of samples. Innovative approaches should be devised to enhance the predictability of the events in a minimal number of samples.

Innovative and interactive data visualization techniques should be appropriately integrated into the data-centric architecture to effectively support the manufacturing users in easily understanding the manufacturing contexts throughout the collected data and the outcomes of prediction models. Human-readable dashboards should represent the critical features of monitored data and the main explanations of the data-driven outcomes to support the decision-making of manufacturing users effectively.

4.1 The SERENA Vision

Looking at a digital future of smart factories embracing the as-a-service model to conduct their business, granting them with resiliency to changes, the SERENA project and its consortium of 14 partners coming from all across Europe, aim to introduce an innovative web platform supporting manufacturers in their maintenance activities. SERENA aims to bring together latest research and development activities under a versatile and scalable system, bridging edge operations to cloud processing.

In particular, the individual objectives of the SERENA integrated system are as follows:

- Collect and process data from different sources and sensors within the on-site and local factory considering edge computing technologies;
- Derive and separate 'smart data' from 'big data';
- Apply advanced data analytics, AI methods and hybrid methods considering both physical models and data driven approaches for predicting potential failures and improve parameters;
- Allow remote access and data processing in cloud for remote prediction and more accurate maintenance actions;
- Enable easy-to-use interfaces for managing data and providing human operator support for machines status and maintenance guidance using AR devices and related high-quality information;
- Provide a security middleware to authorise and secure access to its features.

The outcome should be generic enough to address versatile use cases addressing their deployment requirements and maintenance needs. As a result, the SERENA solutions have been tested and validated in several applications (white goods, metrological engineering, robotics industry, steel parts production, elevators production), also checking the link to other industries (automotive, aerospace etc.) showing the versatile character of the project.

More information are provided:

- in Chapter "A hybrid Cloud-to-Edge Predictive Maintenance Platform", about predictive maintenance architectures and the one proposed in SERENA along with its platform core functionalities and potential deployment.
- in Chapter "Data-Driven Predictive Maintenance: A Methodology Primer", predictive analytics techniques emerged out of SERENA are discussed including data driven and physics based analytics, regarding cloud and edge processing.
- in Chapter "Services to Facilitate Predictive Maintenance in Industry 4.0" the services wrapping the SERENA platform facilitating predictive its users in their predictive maintenance activities.
- in Chapters "Predictive Analytics in Robotic Industry" to "Predictive Maintenance in the Production of Steel Bars: A Data-Driven Approach", the use cases where the SERENA solutions have been deployed and tested are described along with their main outcomes.

- in Chapter "In-Situ Monitoring of Additive Manufacturing", looking on how predictive analytics can find wider adoption, a use case on additive manufacturing is presented along with the application of data-driven analytics.

Acknowledgements This research has been partially funded by the European project "SERENA—VerSatilE plug-and-play platform enabling REmote predictive mainteNAnce" (Grant Agreement: 767561).

References

1. McKinsey & Company, Fourth industrial revolution: beacons of technology and innovation in manufacturing, in *World Economic Forum Annual Meeting* (2019)
2. N.J. Boughton, I.C. ArokiamFirst, The application of cellular manufacturing: a regional small to medium enterprise perspective. Proc. Inst. Mech. Eng. Part B: J. Eng. Manuf. **214**, 751 (2000)
3. Y. Xiang, Z. Zhu, D.W. Coit, Q. Feng, Condition-based maintenance under performance-based contracting, Comput. Ind. Eng. **111**(C), 391 (2017)
4. P. Mehta, A. Werner, L. Mears, Condition based maintenance-systems integration and intelligence using Bayesian classification and sensor fusion. J. Intell. Manuf. **26**(2), 331 (2015)
5. S. Spiegel, F. Mueller, D. Wiesmann, J. Bird, Cost-sensitive learning for predictive maintenance (2018)
6. M. Soliman, Machine reliability and condition monitoring: a comprehensive guide to predictive maintenance planning, Book, ISBN-13 pp. 979–8557261,906 (2020)
7. J. Huang, J.X. You, H.C. Liu, M.S. Song, Failure mode and effect analysis improvement: a systematic literature review and future research agenda. Reliab. Eng. Syst. Safet **199**, (2020)
8. C. et al, The SERENA European Project. https://serena-project.eu/ (2021). Accessed 05 Jan 2021
9. G. Zou, K. Banisoleiman, A. Gonzźlez, M.H. Faber, Probabilistic investigations into the value of information: a comparison of condition-based and time-based maintenance strategies. Ocean Eng. **188**, (2019). https://doi.org/10.1016/j.oceaneng.2019.106181. https://www.sciencedirect.com/science/article/pii/S0029801819303567
10. O. Avalos-Rosales, F. Angel-Bello, A. Álvarez, Y. Cardona-Valdés, Including preventive maintenance activities in an unrelated parallel machine environment with dependent setup times. Comput. Ind Eng **123**, 364 (2018)
11. P. Rokhforoz, B. Gjorgiev, G. Sansavini, O. Fink, Multi-agent maintenance scheduling based on the coordination between central operator and decentralized producers in an electricity market. Reliab. Eng. Syst. Saf. **107495**, (2021)
12. S. Werbińska-Wojciechowska, Delay-time-based maintenance modeling for technical systems–theory and practice. Adv Syst. Reliab. Eng. pp. 1–42 (2019)
13. M. Ghaleb, S. Taghipour, M. Sharifi, H. Zolfagharinia, Integrated production and maintenance scheduling for a single degrading machine with deterioration-based failures. Comp. Ind. Eng. **143**, (2020)
14. K. Hendrickx, W. Meert, Y. Mollet, J. Gyselinck, B. Cornelis, K. Gryllias, J. Davis, A general anomaly detection framework for fleet-based condition monitoring of machines. Mech. Syst. Sig. Proc. **139**, (2020)
15. K. Mykoniatis, A real-time condition monitoring and maintenance management system for low voltage industrial motors using internet-of-things. Proc. Manuf. **42**, 450 (2020)
16. T. Mohanraj, S. Shankar, R. Rajasekar, N. Sakthivel, A. Pramanik, Tool condition monitoring techniques in milling process-a review. J. Mater. Res. Technol. **9**(1), 1032 (2020)
17. L. He, M. Xue, B. Gu, Internet-of-things enabled supply chain planning and coordination with big data services: certain theoretic implications. J. Manage. Sci. Eng. **5**(1), 1 (2020)

18. S. Bader, T. Barth, P. Krohn, R. Ruchser, L. Storch, L. Wagner, S. Findeisen, B. Pokorni, A. Braun, P. Ohlhausen, et al., Agile shopfloor organization design for industry 4.0 manufacturing. Proc. Manuf. **39**, 756 (2019)
19. J. Wang, W. Zhang, Y. Shi, S. Duan, J. Liu, Industrial big data analytics: challenges, methodologies, and applications, arXiv preprint arXiv:1807.01016 (2018)
20. A.C. Márquez, A. de la Fuente Carmona, J.A. Marcos, J. Navarro, Designing cbm plans, based on predictive analytics and big data tools, for train wheel bearings. Comput. Ind. **122**, 103292 (2020)
21. M.C.M. Oo, T. Thein, *An efficient predictive analytics system for high dimensional big data* (J. King Saud Univ.-Comput. Inf, Sci, 2019)
22. X.S. Si, W. Wang, C.H. Hu, D.H. Zhou, Remaining useful life estimation-a review on the statistical data driven approaches. Eur. J. Operat. Res. **213**(1), 1 (2011)
23. P. Dangeti, *Statistics for Machine Learning* (Packt Publishing Ltd. 2017)
24. O. Bousquet, S. Boucheron, G. Lugosi, Introduction to statistical learning theory, in *Summer School on Machine Learning* (Springer, 2003), pp. 169–207
25. R. Gao, L. Wang, R. Teti, D. Dornfeld, S. Kumara, M. Mori, M. Helu, Cloud-enabled prognosis for manufacturing. CIRP annals **64**(2), 749 (2015)
26. T. Zonta, C.A. da Costa, R. da Rosa Righi, M.J. de Lima, E.S. da Trindade, G.P. Li, Predictive maintenance in the industry 4.0: a systematic literature review. Comput. Ind. Eng. 106889 (2020)
27. T.P. Carvalho, F.A. Soares, R. Vita, R.d.P. Francisco, J.P. Basto, S.G. Alcalá, A systematic literature review of machine learning methods applied to predictive maintenance. Comput. Ind. Eng. **137**, 106024 (2019)
28. C.M. Carbery, R. Woods, A.H. Marshall, A bayesian network based learning system for modelling faults in large-scale manufacturing, in *2018 IEEE International Conference on industrial technology (ICIT)* (IEEE, 2018), pp. 1357–1362
29. Y. Guo, Y. Zhou, Z. Zhang, Fault diagnosis of multi-channel data by the cnn with the multilinear principal component analysis. Measurement **171**, (2021)
30. A. Malhi, R. Yan, R.X. Gao, Prognosis of defect propagation based on recurrent neural networks. IEEE Trans. Inst. Measur. **60**(3), 703 (2011)
31. J. Zhang, P. Wang, R. Yan, R.X. Gao, Deep learning for improved system remaining life prediction. Procedia CIRP **72**, 1033 (2018)
32. A.K. Rout, P. Dash, R. Dash, R. Bisoi, Forecasting financial time series using a low complexity recurrent neural network and evolutionary learning approach. J. King Saud Univ.-Comput. Inf. Sci. **29**(4), 536 (2017)
33. H. Yan, Y. Qin, S. Xiang, Y. Wang, H. Chen, Long-term gear life prediction based on ordered neurons LSTM neural networks. Measurement **165**, (2020)
34. M. Sayah, D. Guebli, Z. Al Masry, N. Zerhouni, Robustness testing framework for rul prediction deep lstm networks, ISA transactions (2020)
35. W. Luo, T. Hu, Y. Ye, C. Zhang, Y. Wei, A hybrid predictive maintenance approach for CNC machine tool driven by digital twin. Robot. Comp.-Integ. Manufact. **65** (2020)
36. X. Bampoula, G. Siaterlis, N. Nikolakis, K. Alexopoulos, A deep learning model for predictive maintenance in cyber-physical production systems using lstm autoencoders, Sensors **21**(3) (2021). https://doi.org/10.3390/s21030972. https://www.mdpi.com/1424-8220/21/3/972
37. A. Vaswani, N. Shazeer, N. Parmar, J. Uszkoreit, L. Jones, A.N. Gomez, L. Kaiser, I. Polosukhin, Attention is all you need, arXiv preprint arXiv:1706.03762 (2017)
38. N. Wu, B. Green, X. Ben, S. O'Banion, Deep transformer models for time series forecasting: The influenza prevalence case, arXiv preprint arXiv:2001.08317 (2020)
39. S. Jaskó, A. Skrop, T. Holczinger, T. Chován, J. Abonyi, Development of manufacturing execution systems in accordance with industry 4.0 requirements: A review of standard- and ontology-based methodologies and tools, Computers in Industry **123**, 103300 (2020). https://doi.org/10.1016/j.compind.2020.103300. http://www.sciencedirect.com/science/article/pii/S0166361520305340

40. R.S. Peres, A. Dionisio Rocha, P. Leitao, J. Barata, Idarts – towards intelligent data analysis and real-time supervision for industry 4.0, Computers in Industry **101**, 138 (2018). https://doi.org/10.1016/j.compind.2018.07.004. http://www.sciencedirect.com/science/article/pii/S0166361517306759

41. R.F. Babiceanu, R. Seker, Big data and virtualization for manufacturing cyber-physical systems: A survey of the current status and future outlook, Computers in Industry **81**, 128 (2016). Emerging ICT concepts for smart, safe and sustainable industrial systems

42. I. Simonis, Container-based architecture to optimize the integration of microservices into cloud-based data-intensive application scenarios, in *Proceedings of the 12th European Conference on Software Architecture: Companion Proceedings* (ACM, 2018), p. 34

43. A. Acquaviva, D. Apiletti, A. Attanasio, E. Baralis, L. Bottaccioli, T. Cerquitelli, S. Chiusano, E. Macii, Patti, Forecasting heating consumption in buildings: a scalable full-stack distributed engine, Electronics **8**(5), 491 (2019). https://doi.org/10.3390/electronics8050491

44. B. Xu, S.A. Kumar, Big data analytics framework for system health monitoring, in *2015 IEEE International Congress on Big Data* (2015), pp. 401–408. http://orcid.org/10.1109/BigDataCongress.2015.66

45. G.M. D'silva, A. Khan, Gaurav, S. Bari, Real-time processing of iot events with historic data using apache kafka and apache spark with dashing framework, in *2017 2nd IEEE International Conference on Recent Trends in Electronics, Information Communication Technology (RTEICT)* (2017), pp. 1804–1809. http://orcid.org/10.1109/RTEICT.2017.8256910

46. D. Apiletti, C. Barberis, T. Cerquitelli, A. Macii, E. Macii, M. Poncino, F. Ventura, istep, an integrated self-tuning engine for predictive maintenance in industry 4.0, in *IEEE International Conference on Parallel & Distributed Processing with Applications, Ubiquitous Computing & Communications, Big Data & Cloud Computing, Social Computing & Networking, Sustainable Computing & Communications, ISPA/IUCC/BDCloud/SocialCom/SustainCom 2018, Melbourne, Australia, December 11-13, 2018* (2018), pp. 924–931

47. S. Panicucci, N. Nikolakis, T. Cerquitelli, F. Ventura, S. Proto, E. Macii, S. Makris, D. Bowden, P. Becker, N. O'Mahony, L. Morabito, C. Napione, A. Marguglio, G. Coppo, S. Andolina, A cloud-to-edge approach to support predictive analytics in robotics industry. Electronics **9**(3), 492 (2020)

48. M. Ribeiro, K. Grolinger, M.A.M. Capretz, Mlaas: Machine learning as a service, in *2015 IEEE 14th International Conference on Machine Learning and Applications (ICMLA)* (2015), pp. 896–902. https://doi.org/10.1109/ICMLA.2015.152

49. Y. Jin, H. Lee, On-demand computation offloading architecture in fog networks, Electronics **8**(10), 1076 (2019). https://doi.org/10.3390/electronics8101076

50. Y. Yao, Z. Xiao, B. Wang, B. Viswanath, H. Zheng, B.Y. Zhao, Complexity vs. performance: Empirical analysis of machine learning as a service, in *Proceedings of the ACM Internet Measurement Conference (IMC'17)* (London, UK, 2017)

51. N. Nikolakis, R. Senington, K. Sipsas, A. Syberfeldt, S. Makris, On a containerized approach for the dynamic planning and control of a cyber - physical production system, Robotics and Computer-Integrated Manufacturing **64**(December 2019), 101919 (2020). https://doi.org/10.1016/j.rcim.2019.101919

52. J. Murphree, Machine learning anomaly detection in large systems, in *2016 IEEE AUTOTESTCON* (2016), pp. 1–9

53. Y. Chen, F. Zhu, J. Lee, Data quality evaluation and improvement for prognostic modeling using visual assessment based data partitioning method. Computers in Industry **64**(3), 214 (2013)

54. M. Canizo, E. Onieva, A. Conde, S. Charramendieta, S. Trujillo, Real-time predictive maintenance for wind turbines using big data frameworks (2017). https://doi.org/10.1109/ICPHM.2017.7998308

55. J. Dalzochio, R. Kunst, E. Pignaton, A. Binotto, S. Sanyal, J. Favilla, J. Barbosa, Machine learning and reasoning for predictive maintenance in industry 4.0: Current status and challenges, Computers in Industry **123**, 103298 (2020). https://doi.org/10.1016/j.compind.2020.103298. http://www.sciencedirect.com/science/article/pii/S0166361520305327

56. M.A. Djeziri, S. Benmoussa, E. Zio, Review on Health Indices Extraction and Trend Modeling for Remaining Useful Life Estimation (Springer International Publishing. Cham 183–223 (2020). https://doi.org/10.1007/978-3-030-42726-9_8

57. C. Zhang, P. Lim, A.K. Qin, K.C. Tan, Multiobjective deep belief networks ensemble for remaining useful life estimation in prognostics, IEEE Trans. Neural Netw. Learn. Syst. **PP**(99), 1 (2016)

58. S.A. Asmai, A.S.H. Basari, A.S. Shibghatullah, N.K. Ibrahim, B. Hussin, Neural network prognostics model for industrial equipment maintenance, in *2011 11th International Conference on Hybrid Intelligent Systems (HIS)* (2011), pp. 635–640

59. I. Anagiannis, N. Nikolakis, K. Alexopoulos, Energy-based prognosis of the remaining useful life of the coating segments in hot rolling mill. Appl. Sci. (Switzerland) **10**(19), 6827 (2020). https://doi.org/10.3390/app10196827. https://www.mdpi.com/2076-3417/10/19/6827

60. Z. Huang, Z. Xu, X. Ke, W. Wang, Y. Sun, Remaining useful life prediction for an adaptive skew-wiener process model. Mech. Syst. Sig. Proc. **87**, 294 (2017). https://doi.org/10.1016/j.ymssp.2016.10.027. http://www.sciencedirect.com/science/article/pii/S0888327016304423

61. Z. Zhang, X. Si, C. Hu, Y. Lei, Degradation data analysis and remaining useful life estimation: a review on wiener-process-based methods. Eur. J. Operat. Res. **271**(3), 775 (2018). https://doi.org/10.1016/j.ejor.2018.02.033. http://www.sciencedirect.com/science/article/pii/S0377221718301486

62. D. Wang, K.L. Tsui, Brownian motion with adaptive drift for remaining useful life prediction: revisited, mechanical systems and signal processing **99**, 691 (2018). https://doi.org/10.1016/j.ymssp.2017.07.015. http://www.sciencedirect.com/science/article/pii/S0888327017303771

63. B. Wang, Y. Lei, N. Li, N. Li, A hybrid prognostics approach for estimating remaining useful life of rolling element bearings. IEEE Trans. Reliab. **69**(1), 401 (2020). https://doi.org/10.1109/TR.2018.2882682

64. J. Deutsch, D. He, Using deep learning-based approach to predict remaining useful life of rotating components. IEEE Trans. Syst. Man Cybern. Syst. **48**(1), 11 (2018). http://orcid.org/10.1109/TSMC.2017.2697842

65. Y. Chen, G. Peng, Z. Zhu, S. Li, A novel deep learning method based on attention mechanism for bearing remaining useful life prediction. Appl. Soft Comput. **86**, (2020). https://doi.org/10.1016/j.asoc.2019.105919. http://www.sciencedirect.com/science/article/pii/S1568494619307008

66. A. Zhang, H. Wang, S. Li, Y. Cui, Z. Liu, G. Yang, J. Hu, Transfer learning with deep recurrent neural networks for remaining useful life estimation. Appl. Sci. **8**(12) (2018). https://doi.org/10.3390/app8122416. https://www.mdpi.com/2076-3417/8/12/2416

67. X. Li, W. Zhang, Q. Ding, Deep learning-based remaining useful life estimation of bearings using multi-scale feature extraction. Reliab. Eng. Syst. Saf. **182**, 208 (2019). https://doi.org/10.1016/j.ress.2018.11.011. http://www.sciencedirect.com/science/article/pii/S0951832018308299

68. S. Zheng, K. Ristovski, A. Farahat, C. Gupta, Long short-term memory network for remaining useful life estimation, in *2017 IEEE International Conference on Prognostics and Health Management (ICPHM)* (2017), pp. 88–95. https://doi.org/10.1109/ICPHM.2017.7998311

69. J. Liu, Q. Li, W. Chen, Y. Yan, Y. Qiu, T. Cao, Remaining useful life prediction of pemfc based on long short-term memory recurrent neural networks. Int. J. Hydrogen Energy **44**(11), 5470 (2019). https://doi.org/10.1016/j.ijhydene.2018.10.042. http://www.sciencedirect.com/science/article/pii/S0360319918332191. The 6th International Conference on Energy, Engineering and Environmental Engineering

70. H. Liu, Z. Liu, W. Jia, X. Lin, S. Zhang, A novel transformer-based neural network model for tool wear estimation. Measur. Sci. Technol. **31**(6), 065106 (2020). https://doi.org/10.1088/1361-6501/ab7282

71. B. He, L. Liu, D. Zhang, Digital twin-driven remaining useful life prediction for gear performance degradation: a review. J. Comput. Inf. Sci. Eng. 1–70

72. T. Cerquitelli, D.J. Pagliari, A. Calimera, L. Bottaccioli, E. Patti, A. Acquaviva, M. Poncino, Manufacturing as a data-driven practice: methodologies, technologies, and tools. Proc. IEEE **109**(4), 399 (2021). https://doi.org/10.1109/JPROC.2021.3056006

A Hybrid Cloud-to-Edge Predictive Maintenance Platform

Angelo Marguglio, Giuseppe Veneziano, Pietro Greco, Sven Jung,
Robert Siegburg, Robert H. Schmitt, Simone Monaco, Daniele Apiletti,
Nikolaos Nikolakis, Tania Cerquitelli, and Enrico Macii

Abstract The role of maintenance in the industry has been shown to improve companies' productivity and profitability. Industry 4.0 revolutionised this field by exploiting

A. Marguglio · G. Veneziano · P. Greco
Research & Innovation, Engineering Ingegneria Informatica S.p.A., Palermo, Italy
e-mail: angelo.marguglio@eng.it

G. Veneziano
e-mail: giuseppe.veneziano@eng.it

P. Greco
e-mail: pietro.greco@eng.it

S. Jung · R. Siegburg · R. H. Schmitt
Fraunhofer Institute for Production Technology IPT and Laboratory for Machine Tools and
Production Engineering (WZL) of RWTH Aachen University, Aachen, Germany
e-mail: sven.jung@ipt.fraunhofer.de

R. Siegburg
e-mail: robert.siegburg@ipt.fraunhofer.de

R. H. Schmitt
e-mail: robert.Schmitt@ipt.fraunhofer.de

S. Monaco · D. Apiletti · T. Cerquitelli (✉)
Department of Control and Computer Engineering, Politecnico di Torino, Turin, Italy
e-mail: tania.cerquitelli@polito.it

S. Monaco
e-mail: simone.monaco@polito.it

D. Apiletti
e-mail: daniele.apiletti@polito.it

N. Nikolakis
Laboratory for Manufacturing Systems and Automation, University of Patras, Patras, Greece
e-mail: nikolakis@lms.mech.upatras.gr

E. Macii
Interuniversity Department of Regional and Urban Studies and Planning, Politecnico di Torino,
Turin, Italy
e-mail: enrico.macii@polito.it

© Springer Nature Singapore Pte Ltd. 2021
T. Cerquitelli et al. (eds.), *Predictive Maintenance in Smart Factories*,
Information Fusion and Data Science,
https://doi.org/10.1007/978-981-16-2940-2_2

emergent cloud technologies and IoT to enable predictive maintenance. Significant benefits can be obtained by taking advantage of historical data and Industrial IoT streams, combined with high and distributed computing power. Many approaches have been proposed for predictive maintenance solutions in the industry. Typically, the processing and storage of enormous amounts of data can be effectively performed cloud-side (e.g., training complex predictive models), minimising infrastructure costs and maintenance. On the other hand, raw data collected on the shop floor can be successfully processed locally at the edge, without necessarily being transferred to the cloud. In this way, peripheral computational resources are exploited, and network loads are reduced. This work aims to investigate these approaches and integrate the advantages of each solution into a novel flexible ecosystem. As a result, a new unified solution, named SERENA Cloud Platform. The result addresses many challenges of the current state-of-the-art architectures for predictive maintenance, from hybrid cloud-to-edge solutions to intermodal collaboration, heterogeneous data management, services orchestration, and security.

1 Introduction

Predictive Maintenance can reveal useful insights into the manufacturing processes, be they in industrial production monitoring [1, 2] or in related sectors (e.g. Parcel Delivery Services [3]). Significant benefits can be obtained by taking advantage of historical data and Industrial Internet of Things (IIoT) streams combined, thanks to high and distributed computing power. This information can then be exploited by plant managers to anticipate maintenance actions, thus reducing machinery downtimes and costs. To be sufficiently reliable, predictive solutions need to collect and process large amounts of data [4].

Even if the predictive maintenance task is quite straightforward in concept, the actual design and provisioning of a working, reliable and effective solution requires addressing many challenges. For instance, large-scale computing power is typically available in the cloud, whereas data are collected on-premises. Hence, solutions can either adopt a cloud-first approach or an edge-computing approach. The proposed work provides instead a flexible and hybrid solution to meet the real-world industrial requirements. Other challenges are related to heterogeneous data management, due to the multiple sources of data, the orchestration of distributed services, their intermodal collaboration and their security. To these challenges, the proposed work provides an effective solution, by combining and advancing the state of the art with a flexible and holistic approach.

To this aim, the chapter provides the following contributions.

- A literature review of the state of the art, reporting the challenges and principles regarding the design of cloud platforms for predictive maintenance.

- The design choices of the proposed solution, the SERENA Cloud Platform, and its relationships with the approaches found in literature, highlighting differences and similarities.
- The architecture, technologies, and deployment choices of the SERENA Cloud Platform, its components, and its operational contexts.

The chapter is organised as follows. Section 2 presents related works in predictive maintenance, cloud platforms, and their technologies. Section 3 describes the design choices of the SERENA Cloud Platform and the challenges addressed by the proposed solution. Section 4 illustrates its architecture, components, and deployment. Finally, Sect. 5 draws conclusions and provides future work directions.

2 Related Works

2.1 Predictive Maintenance in Industry 4.0

Industry 4.0 has revolutionised manufacturing and brought new digital technologies into the industry that are transforming the production process. The role of maintenance in improving companies' productivity and profitability has been shown in [5], enabled by recent advantages in Information and Communications Technology (ICT). Predictive maintenance can reveal underlying information in large historical data. Differently from traditional industries, companies based on Industry 4.0 change their production process by combining the strengths of optimised manufacturing with Internet technologies, establishing new manufacturing processes, maintenance management and related strategies through the use of Big Data. The key enabler for this revolution is therefore the emergent use of new technologies, such as Cyber-Physical Systems (CPSs) [6], machine-to-machine communication [7], and cloud technologies [8], that form the basis to gather and analyse data across machines.

First approaches on predictive maintenance platforms were implemented and introduced [9, 10], but fail to address the tension between flexibility and providing the required key properties of a distributed system. One promising approach is an integrated platform consisting of three main pillars responsible for dealing with individual aspects [11]. The first pillar is responsible for data extraction and analysis, the second pillar is responsible for maintenance modelling, knowledge modelling and representation, and the third pillar had advisory capabilities on maintenance planning.

As described in [12], the IIoT massively increases the available amount of data, allowing, on one hand, the construction of reliable predictive models, but on the other hand, bearing new challenges for the manufacturing ecosystem. The scale of sophisticated functions and machine learning models for predictive maintenance require powerful and efficient cloud infrastructures.

2.2   Cloud and Edge Computing

Cloud computing [13] allows developing frameworks able to crunch large quantities of data and train powerful models. Such solutions are mostly limited by the amount of data that can be sent to the server over the network to still grant real-time responses. To overcome this limitation, data transmission can be reduced [14]: the data collected from sensors are pre-processed through the use of middleware agents, which reduce the amount of data sent to the server. These solutions move part of the computation to the edge of the network, namely with an edge-computing strategy [15].

Cloud solutions provide centralised storage and processing capacity and thus minimise the cost and restriction for automation and maintenance infrastructure [10]. Advantages like virtualisation, parallel processing and data security, enable the efficient processing of large amounts of data and make cloud solutions suitable for the training of complex predictive models. This centralised approach also comes with a cost as it introduces latency depending on the network topology and bandwidth. In the same way, the response time of analytics-driven decision making is supposed to increase accordingly.

Architectures as the one proposed in [16], focusing on predictive maintenance on medical devices show the benefits of a centralised solution. Moreover, the presented results rely on light-weight learning algorithms (genetic algorithms) tested on a small number of devices. It can be expected that the same approach would face structural limitations if one of these two elements disappear.

A different approach is to extend cloud computing to the edge nodes of a network, the so-called fog computing [17]. Depending on where the intelligence capability is placed when distributing the computational power over the network, we talk about "fog" computing and "edge" computing. Whereas both the approaches involve moving the data processing down to where the information is originating, the former requires a gateway node close to the devices under controls, while in the latter the devices themselves perform the computation.

In both fog and edge distributed architectures, the advantages of centralised cloud computing can be preserved while reducing the needed bandwidth for end-to-end communication towards and between field devices [18]. Following this approach, intelligence features, such as critical functions, are located at an edge node in a factory, reducing the transported amount of data and thus the latency and response time [19].

Nevertheless, distributed architectures also bring in new challenges [20]. From the standardised communication of an inter-nodal collaboration to the IoT requirement of an end-to-end infrastructure with a collaboration between the nodes of the edge and the cloud platform. In fog computing, it is very common that nodes use peer-to-peer communication. By definition, this communication can be omnidirectional, meaning that every node can address every other node regardless of the position in the network and the lack of a single control entity [21]. This brings in a lot of flexibility, but also sensible complexity and management overhead since control and data flows can become nontransparent. This can be drastically reduced by restricting peer-to-peer communication to a hierarchical order, where a manager node controls the communication and the functionalities of underlying worker nodes [22]. Even if this introduces a lot more control and structure, in fog networks with a functional decomposition into multiple computing nodes, this could quickly become a limitation due to increased bandwidth utilisation and latency. To overcome this challenge, message brokers have been introduced, which act as an intermediate for messages and enable a one-to-many message distribution and a decoupling of applications [23]. This approach focuses on delivering messages from producers to consumers. However, the centralised approach brings in the drawback of a single point of failure.

Another challenge is the uniform data handling and data quality among the whole system. As data form the basis for predictive maintenance methods, efficient handling and management are crucial. Storage capacities of distributed nodes not only may be limited, but data also need to be globally available as input for analytical and control functionalities. This demands efficient data storage solutions and query interfaces. Even if there are many different types of databases available depending on the types of data to be stored, like file-based, SQL-based, graph-based, NoSQL-based, in order to overcome this, different approaches propose central injection services coordinating data access among various nodes and database types or distributed database systems [24]. Besides data storage, a big challenge is to uniformly handle data coming from heterogeneous data sources, which may appear in different formats and sampling rates [25]. Furthermore, in order to be able to correlate machine conditions with process information, the data need to be semantically augmented. Thus, also semantic frameworks need to be introduced.

Asset management and orchestration, due to the modular nature of distributed applications, causes individual functionalities to be distributed among different nodes in the network. Thus, the coordination of these resources and functionalities is essential [20]. On the one hand, applications have to be managed and developed for heterogeneous-distributed devices. Programming platforms provide a simplified abstraction of programming models and a common basis for a heterogeneous deployment [26]. On the other hand, resources and efficient deployment have to be handled. QoS-aware self-adaptive schedulers are able to automatically scale applications and plan re-configurations and placement of resources [27], like offloading computational tasks from resource constraint devices to more powerful devices for execution or deploying functionalities to the required location in the edge of the network.

In distributed systems, security concerns are quite high, because they relate to the distributed architecture and functionalities between nodes and cloud systems. Within this context, unwanted access and behaviour in the ecosystem have to be prevented [28]. Distributed systems process sensitive data coming from devices and sensors and therefore should consider proper privacy assurance. To address this, a suitable authentication and policy management needs to manage the access within the ecosystem, which can be further distinguished between user authentication, device authentication, data migration authentication and instance authentication. Additionally, encryption methods enable a secure data exchange between nodes of the network.

2.3 Predictive Maintenance Technologies

Predictive Maintenance approaches can be organised in different categories, depending on the type of sensors applied to the monitored devices [29]. Such type of data can be passively originated from process sensors, already available inside machines, or from test sensors applied to detect anomalies in the component movements (e.g. accelerometers to measure irregular vibrations). A different category involves a collection of information provided by injecting test signals into the equipment and measuring the response. All these possibilities have been used in different solutions in order to produce the machine-state prediction, handling different issues.

In [5] various transparent predictive approaches have been integrated to provide a scalable predictive-maintenance solution with a focus on end-user interpretability. Healthcare and living environments are targeted as applications of Internet of Things solutions in [8], which provides various open technological research directions.

In [19] Big Data processing techniques for energy-related applications are presented, together with their relationships with manufacturing and Internet of Things solutions.

Communication issues related to the collection of remote data and the corresponding edge solutions are presented in [20] and in [22], with the latter providing a focus on Big Data spatial solutions based on micro-service technologies deployed at the edge.

Different approaches are known in the literature to design and implement intermediate concepts to build platforms supporting predictive maintenance. The work proposed in this chapter aims to integrate the advantages of different solutions into a unique ecosystem, SERENA, to meet the real-world requirements coming from industrial usage. The SERENA solution is a versatile cloud platform, capable of adopting a flexible hybrid edge-cloud deployment model and seamlessly incorporate software and hardware components, dynamically deployed both at the plant and in the cloud. Besides its dynamic flexibility, SERENA, preliminary presented in [30], addresses current limitations and challenges by providing the following contributions: (i) intermodal collaboration among nodes is provided by means of a scalable message broker, (ii) heterogeneous data management is provided by a specifically designed MIMOSA-based data model, (iii) services orchestration is based on state-of-the-art containerised solutions, and (iv) security is provided by a custom component named Reverse Proxy Certification Authority (RPCA). SERENA contributions on each aspect are described in Sect. 3.

3 SERENA Design

SERENA leverages well-established enterprise-grade technologies, such as Docker and its cluster orchestrator Docker Swarm, as well as other Big Data technologies, enabling a flexible solution that can be tailored to different real-world use cases. The

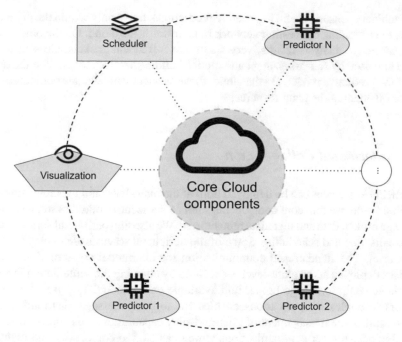

Fig. 1 Services can be classified into core components and peripheral services, at the edge

main design choices that make SERENA a valuable contribution, addressing current limitations in the state of the art, are provided in the following.

SERENA has been designed by combining services supporting predictive maintenance both at the core and at the edge, as exemplified in Fig. 1, where the latter can be deployed both in the cloud or on-premises, i.e., inside the manufacturing plant. Examples of peripheral services are predictors, whereas storage and high-performance processing components are typical examples of core services.

Moreover, thanks to the plug-and-play capabilities of SERENA, peripheral services can be added to the platform in an on-demand fashion, and with minimal effort.

Being the core components deployed on the cloud, cost and restrictions for infrastructure maintenance are minimised [10]. Moreover, having the core components on the same farm, ensures that communication among them is fast and easier to secure. Indeed, the services running within Docker Swarm are connected by an isolated and dedicated docker network layer that does not publish ports out of the farm. Only a scalable reverse proxy acts as a front-end encrypting, filtering and routing all the traffic exchanged with the external world. However, when the number of plant-level devices subject to predictive analysis increases, the traffic between edge and cloud could overload the network and affect the system performance. In SERENA this issue has been overcome thanks to the flexibility of the components' deployment model. Thus, the predictive analysis service can run either at the plant [19] or cloud side. When predictors run plant-side, heavy and frequent packets of data produced by machinery sensors, useful to generate predictions, travel only within the plant net-work, thus avoiding congesting network links towards the cloud. On the other hand, only predictive analysis results, very small and infrequent packets, are sent to the cloud and used by self-assessment and model-building services to improve the qual-ity of the predictive model. On the cloud, these services can leverage computational power off-loading the plant resources.

3.1 Intermodal Collaboration

In SERENA, services can be deployed both at the plant-level and the cloud side. On the cloud-side, we find core components such as the security middleware, a scalable message broker, data and metadata repositories. We also find peripheral services such as visualisation and scheduling. Some of the peripheral services are deployed at the plant-level. The standardised communication for intermodal collaboration among services deployed at the plant-level is executed by allowing their interaction through a scalable central message broker hub available on the cloud. The use of message brokers decouples the applications, enables one-to-many messages distribution [23], and simplifies the management of plug-and-play capabilities. This approach allows avoiding peer-to-peer communications which can lead to complexity and manage-ment overhead.

3.2 Data Management

In SERENA, sensor data coming from the plant are sent to the central message broker, where data are indexed and stored along with metadata. Since data reposi-tories are shared, direct communication among all the SERENA services becomes unnecessary. Central injection services are used to coordinate access to data com-ing from various nodes regardless of their location. A purposefully designed data model based on the MIMOSA standard is used to uniformly handle data coming from heterogeneous data sources. MIMOSA is an open standard for several key asset management areas, including those related to factories. In particular, sensors' data, fetched from heterogeneous legacy systems, are normalised by ad-hoc software middleware solutions and represented in JSON-LD messages along with metadata, defining how to interpret them. In this way, all the services that need to access data can always expect the same format and information on how to use them. In this way, many limitations, such as those stated in [25], can be overcome. The proposed data model also associates metadata with physical assets, measurements coming from the plant, and predictors' outcomes. Therefore, data are semantically augmented to correlate machine conditions with process information.

3.3 Service Orchestration

Due to the distributed nature of solutions for predictive maintenance, individual services might have to be distributed on heterogeneous nodes, which may differ by hardware or software configurations. In SERENA, these concerns have been addressed by using Docker, which allows deploying applications regardless of the implementation technology and characteristics of the target deployment environment. In addition, whenever services are deployed as part of a Docker Swarm, their management is simplified thanks to the primitives offered by Docker. For instance, Docker allows to easily scale a service by changing the number of replicas.

3.4 Security

In a distributed system, where data flows from plant-level nodes to the cloud over an untrusted network, security is a major concern. In particular, the main challenges are to prevent unwanted access and malicious behaviour to the cloud [28], to assure the privacy of sensitive data coming from devices and sensors and finally, to authenticate access to the system, both for users and devices. In SERENA, these concerns have been addressed through a custom component named Reverse Proxy Certification Authority (RPCA). It provides a two-way authentication mechanism for both users and plant-level devices through TLS certificates. As the challenge is still to manage and automate the distribution of these certificates, a certificate management and distribution protocol has been designed and implemented on top of the RPCA, as well as software components to decrypt messages carrying certificates. In addition, communication between core services within the cloud or the Docker Swarm can be further strengthened using additional data encryption services offered by Docker.

4 SERENA Architecture

The SERENA platform, briefly described in [31], includes several services for effective and efficient predictive maintenance. Its architecture is based on a light-weight micro-services architecture, exploiting Docker containers to wrap the services into flexible units.

The strategy of wrapping services in containers isolates them from the underlying host infrastructure. Thus it allows the services composing the SERENA system to be deployed on a wide variety of infrastructures, from hardware servers and gateways, through virtual machines, to hosted environments on public and hybrid clouds. Whilst the SERENA reference implementation uses specific technology to realise each service, the common interface allows technology to be swapped, depending on the specific implementation requirements. This technology transparency is an important concept in SERENA's plug-n-play architecture providing scalability and modularity. The instantiation and life cycle of the services are managed by the Docker Swarm orchestrator, which provides a series of functionalities, such as scaling, service discovery and load balancing, easing the distribution of services across the cloud nodes. Moreover, it ensures that services are automatically redeployed in case any of them fails. The required Docker Images are downloaded by the cloud nodes from a Local Docker Registry in which images are stored after undergoing a validation process. This process aims to ensure that no security vulnerabilities are introduced in the production environment. Finally, the Docker orchestration layer also provides an overlay network that allows the services to transparently communicate, irrespective of where they are deployed.

As mentioned before, the SERENA services can be split up into two subsets, the core cloud and peripheral services (Fig. 1).

The core cloud components constitute the essential functionality of the SERENA platform, data and metadata ingestion, digestion and management. These components, still loosely coupled, provide the peripheral services with the functionalities needed to take full advantage of the resources made available by the platform, data coming from the plants and related metadata. Data are measurements gathered from the plant-level machinery and brought to the SERENA cloud through the so-called gateways. Collected data are enriched with metadata, both identifying the source and providing other information, such as physical quantities, measurement units and data types otherwise not tracked. In Fig. 2, the services, their mutual relationships and the overall data flow are illustrated.

A gateway is a software and/or hardware device whose task is to interact with factory sensors and other legacy facilities to fetch measurements data coming from plant-level machinery. Fetched data are then encapsulated in JSON-LD messages along with metadata, and sent to the cloud ingestion layer. There, the RPCA Security Manager, leveraging a Transport Layer Security (TLS) two-way authentication, determines if the incoming messages are originated by a legitimate source. Accepted messages are forwarded to the Central Message Broker which is in turn in charge of dispatching them to the other core cloud components. In particular, the JSON-LD messages are stored in Hadoop and the extracted metadata is sent to the metadata service. The stored data and metadata are made available to the peripheral services through a set of RESTful and RPCA endpoints. The SERENA Platform comprises several predictive analytics, visualisation and scheduling peripheral services. Moreover, any number of additional services can be added, updated and removed with minimal effort.

4.1 Core Services

This section describes the SERENA Cloud Core components that implement the SERENA system base functionality leveraging the plug-and-play capabilities offered by the underlying Docker Swarm technology.

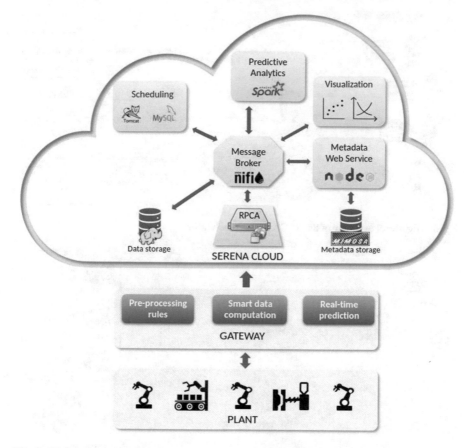

Fig. 2 High-level architecture overview with services classification

4.1.1 RPCA: Reverse Proxy Certification Authority

Security in communication channels in the SERENA platform is of primary importance. Thus, the ones existing between the clients external to the cloud and the cloud itself must be secured against interception and tampering. It must be ensured that any client—hardware device (i.e. plant-level equipment capable of sending data to the cloud), the user or the Docker host (i.e. computer running a Docker daemon)—interacting with the platform cannot be spoofed. Then these entities are authenticated and their access to the platform resources regulated. To this end, the RPCA component, depicted in Fig. 3, acting as a front-end for all the requests to the cloud, implements a two-way TLS authentication mechanism to reject the requests coming from untrusted clients and to forward the legitimate ones to other services inside the cloud. Leveraging open-source technologies, it implements the SERENA security middleware.

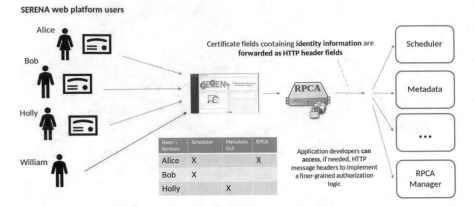

Fig. 3 High-level overview of the RPCA access mechanism

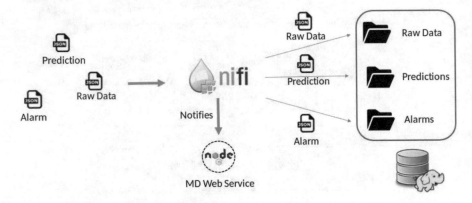

Fig. 4 High-level view of NiFi central message broker functioning

4.1.2 Central Message Broker Service

The central message broker service is the main communication hub of the SERENA system, all the incoming messages, as well as the fetching requests for data or metadata, are dispatched to appropriate different core cloud components. The SERENA message broker is implemented in Apache NiFi: an easy to use, powerful, and reliable system to process and distribute data through scalable directed graphs of data routing, transformation, and system mediation logic. It is presented in Fig. 4. These graphs, called Flows, are composed of a set of configurable processors which wrap the logic needed to interact with other subsystems.

4.1.3 Data Store

The SERENA platform needed an infrastructure able to store and process a huge volume of semi-structured data, related to measurements collected by the gateways from the plant-level equipment or predictions returned by the analytics services. To this end, the open-source framework Apache Hadoop has been employed. It allows the distributed processing of large datasets across clusters of computers running on commodity hardware, providing highly fault-tolerant and high throughput access to data through its distributed filesystem (HDFS). Leveraging a cluster-in-a-cluster model of deployment, Hadoop has been implemented on top of Docker Swarm. Running the data nodes or name nodes as any other SERENA service, their number can be varied by instantiating/de-instantiating the corresponding Docker services. Multiple Hadoop nodes increase the storage capacity and share the workload to efficiently handle storage requests coming from several injectors.

4.1.4 Metadata Service

The Metadata Service was developed for the SERENA European Union H2020 Project. Metadata provides the context and information on the sensor and smart data collected from the gateways. Then it summarises basic information about data, which can make locating and understanding the data more effectively.

MIMOSA OSA-EAI provides open data standards in several key asset management areas:

1. Asset register management
2. Work management
3. Diagnostic and prognostic assessment
4. Vibration and sound data
5. Oil, fluid and gas data
6. Thermographic data
7. Reliability information.

These seven areas are defined by a relational model named Common Relational Information Schema (CRIS). The CRIS defines asset management entities, their attributes and associated types, and relationships between entities. MIMOSA consists of over 400 interconnected tables, which cover many aspects of factory asset management. Figure 5 shows the MIMOSA OSA-EAI layer. A reference data library sits on top of the CRIS. The library contains reference data compiled by MIMOSA, which can be stored by the CRIS.

The data and metadata have to be passed from the edge gateway to the SERENA cloud. For this, we need a message format. SERENA uses JSON-LD as the message format to pass the data and metadata. JSON-LD is a self-describing message structure with data and semantic context for the data. It is an extension to the standard JSON message format, which supports Linked Data.

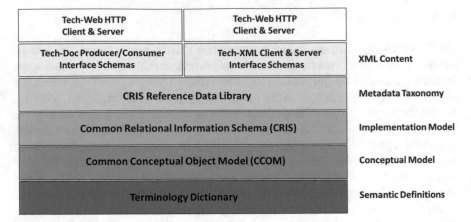

Fig. 5 MIMOSA OSA-EAI layer

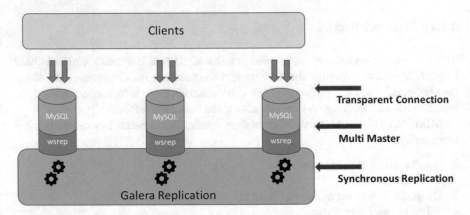

Fig. 6 Galera cluster

JSON-LD and the SERENA vocabulary based on MIMOSA is SERENA's canonical data model, which facilitates seamless communication between all the components in SERENA.

Metadata repository stores information related to the manufacturing plant, its equipment and how the sensor and smart data is collected and stored. The metadata repository consists of a relational database since MIMOSA is a relational schema. The raw data from edge gateways are stored in HDFS and the metadata is stored in the database, which also stores and cross-references the predictions from analytics service.

A metadata repository cluster is used in SERENA for high-availability, which provides high system up-time, no data loss and scalability. It is based on the official version of the MariaDB Galera cluster database, whose architecture is presented in Fig. 6.

The metadata web service, which is implemented in NodeJS, provides a consistent and technology-neutral interface to the metadata repository from other SERENA services. The interface is a set of REST APIs that allow other SERENA services to query and update the metadata repository—GET retrieves the metadata from the metadata repository in JSON-LD format.

4.2 Peripheral Services

Peripheral Services are orchestrated so that they can be automatically deployed to the edge. Model retraining is performed in the cloud, then gateways are updated with the new dockerized models. The orchestrator also allows customizing the particular production process targeted by the gateway models, by means of function labels, whose purpose is to determined the correct service deployment.

- Analytics Services (Model-Building, Prediction services, Self-Assessment)
- Scheduler
- Visualization

The Model building service uses measurement data stored in HDFS to build a prediction model through machine learning algorithms. These algorithms leverage the deployed Python libraries (NumPy, Pandas, SciPy and Scikit-learn) offered by the Analytics Framework. Through Apache Spark, the required data can be fetched either by accessing directly HDFS on-demand or by monitoring in real-time the desired folders. The Prediction services, by using the previously built Model and the Streaming module of Spark, can process real-time data streams to identify possible failures. Predictions are then sent to the Central Message Broker which will store them into HDFS notifying the Metadata Service for indexing purposes. The outcomes of the Predictors are finally analysed by the Self-assessment service to tune, if needed, the prediction Model.

The Scheduling application, instead, aims to optimise the automated planning of the maintenance activities of the shop floor. By retrieving information regarding the remaining useful life of the machines from the analytics modules, the Scheduling application can generate a maintenance schedule that will satisfy several user-defined criteria and will also respect resources availability as well as maintenance task precedence constraints.

The Visualisation application allows to exploit the generated maintenance workflows, providing plotting capabilities, alerting messages based on predictions and Augmented Reality sub-services through which, taking full advantage of smart devices, the workflows are notified to the plant operator.

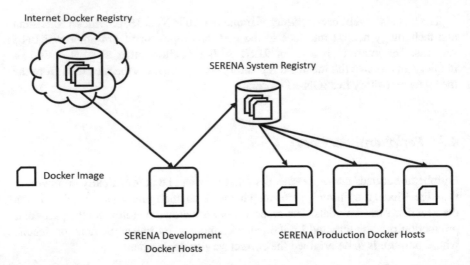

Fig. 7 SERENA provisioning and deployment environment

4.3 Provisioning and Deployment

The SERENA system implements its own local image registry where the service Docker images are stored, as reported in Fig. 7. Before being deployed on the production Docker hosts, the Docker images used to implement the SERENA services have to be rigorously tested and validated for vulnerabilities. This procedure is carried out on development hosts (separated from the production hosts) ensuring that vulnerabilities are not introduced in the platform environment.

After completing the testing procedure, the images are pushed to the Local Docker Registry. The SERENA docker production hosts can access validated images only from this registry. Moreover, to increase the level of security, the TLS two-way authentication is used between the Local Docker Registry and each Docker Production Host so that the authenticity of the source can be trusted. The needed TLS certificates are generated by the previously described RPCA Security Manager component.

5 Conclusions

This chapter presented the SERENA cloud platform, a solution capable of fulfilling the requirements of predictive maintenance analytics in real-world industrial settings by combining a variety of advantages.

The contributions of the proposed platform are wide across the technology stack, data management, and service provisioning and orchestration. Edge-side gateways have been developed to interact with legacy factory data-gathering systems that otherwise could not have been used to bring data to the cloud. Metadata management functionalities are offered so that updates made by the users to metadata through the web interfaces can be distributed to the gateways all over the plants easily. The Central Message Broker, the Metadata Service and Hadoop implementation of the core cloud functionalities can provide reliable and easy-to-use data access interfaces to the pluggable peripheral services. In particular, being the core cloud services implemented as containerised clusters, they assure data redundancy, load balancing and high availability, thus constituting a valid solution for high data volumes. The underlying shared storage has been made available in two versions: one based on the classic NFS and another based on the more sophisticated Dell Technologies Elastic Cloud Storage (ECS) solution. The first one allows a lightweight standard solution, whereas the second one provides an increased level of resilience, scalability and maintainability. The RPCA service, by implementing a flexible, extended form of two-way authentication, secures the communications between an external client and the cloud environment. The designed mechanism has been exploited not only for the gateway-cloud channel but also for the requests made by an operator to access, via browser, all the web service interfaces. Finally, the Data Analytics Framework allowed the deployment of fundamental components such as the predictors, providing them with a complete Python Scientific Stack and with all the facilities needed to efficiently and effectively access the data assets stored in Hadoop. The architecture proposed in this chapter has been tested in different real-world application contexts, ranging from robot companies (Chap. 5) to production of steel bars (Chap. 9), including the metrology engineering industry (Chap. 6), white goods industry (Chap. 7), and the production of elevators (Chap. 8).

Among the different future research directions, we envisage the introduction of web abstractions that can facilitate the use of cloud components. Furthermore, the optimisation of the hardware resources allocation, by means of machine learning models, is additional future work.

Acknowledgements This research has been partially funded by the European project SERENA VerSatilE plug-and-play platform enabling REmote predictive mainteNAnce (Grant Agreement: 767561).

References

1. D. Apiletti, C. Barberis, T. Cerquitelli, A. Macii, E. Macii, M. Poncino, F. Ventura, istep, an integrated self-tuning engine for predictive maintenance in industry 4.0, in *IEEE International Conference on Parallel & Distributed Processing with Applications, Ubiquitous Computing & Communications, Big Data & Cloud Computing, Social Computing & Networking, Sustainable Computing & Communications, ISPA/IUCC/BDCloud/SocialCom/SustainCom 2018, Melbourne, Australia, December 11-13, 2018*, ed. by J. Chen, L.T. Yang (IEEE, 2018), pp. 924–931
2. S. Proto, F. Ventura, D. Apiletti, T. Cerquitelli, E. Baralis, E. Macii, A. Macii, Premises, a scalable data-driven service to predict alarms in slowly-degrading multi-cycle industrial

processes, in *2019 IEEE International Congress on Big Data, BigData Congress 2019, Milan, Italy, July 8-13, 2019*, ed. by E. Bertino, C.K. Chang, P. Chen, E. Damiani, M. Goul, K. Oyama (IEEE, 2019), pp. 139–143

3. S. Proto, E.D. Corso, D. Apiletti, L. Cagliero, T. Cerquitelli, G. Malnati, D. Mazzucchi, Redtag: A predictive maintenance framework for parcel delivery services. IEEE Access **8**, 14953 (2020)

4. T. Cerquitelli, D.J. Pagliari, A. Calimera, L. Bottaccioli, E. Patti, A. Acquaviva, M. Poncino, Manufacturing as a data-driven practice: methodologies, technologies, and tools. Proc. IEEE (2021)

5. I. Alsyouf, The role of maintenance in improving companies' productivity and profitability. Int. J. Prod. Econom. **105**(1), 70 (2007). https://doi.org/10.1016/j.ijpe.2004.06.057

6. E.A. Lee, The past, present and future of cyber-physical systems: a focus on models. Sensors (Switzerland) **15**(3), 4837 (2015). https://doi.org/10.3390/s150304837

7. D. Gorecky, M. Schmitt, M. Loskyll, D. Zühlke, Human-machine-interaction in the industry 4.0 era, in *2014 12th IEEE International Conference on Industrial Informatics (INDIN)* (IEEE, 2014), pp. 289–294

8. S. Wang, Z. Liu, Q. Sun, H. Zou, F. Yang, Towards an accurate evaluation of quality of cloud service in service-oriented cloud computing. J. Intell. Manuf. **25**(2), 283 (2014)

9. J. Lindström, H. Larsson, M. Jonsson, E. Lejon, Towards intelligent and sustainable production: combining and integrating online predictive maintenance and continuous quality control. Proced. CIRP **63**, 443 (2017)

10. L. Spendla, M. Kebisek, P. Tanuska, L. Hrcka, Concept of predictive maintenance of production systems in accordance with industry 4.0, in *2017 IEEE 15Th International Symposium on Applied Machine Intelligence and Informatics (SAMI)* (IEEE, 2017), pp. 000,405–000,410

11. K. Efthymiou, N. Papakostas, D. Mourtzis, G. Chryssolouris, On a predictive maintenance platform for production systems. Proced. CIRP **3**, 221 (2012)

12. Z. Li, Y. Wang, K.S. Wang, Intelligent predictive maintenance for fault diagnosis and prognosis in machine centers: industry 4.0 scenario, Adv. Manuf. **5**(4), 377 (2017)

13. B. Schmidt, L. Wang, Cloud-enhanced predictive maintenance. Int. J. Adv. Manuf. Technol. **99**(1), 5 (2018)

14. J. Wang, L. Zhang, L. Duan, R.X. Gao, A new paradigm of cloud-based predictive maintenance for intelligent manufacturing. J. Intell. Manuf. **28**(5), 1125 (2017)

15. W.Z. Khan, E. Ahmed, S. Hakak, I. Yaqoob, A. Ahmed, Edge computing: a survey. Future Generat. Comput. Syst. **97**, 219 (2019)

16. L.S. Terrissa, S. Meraghni, Z. Bouzidi, N. Zerhouni, A new approach of phm as a service in cloud computing, in *2016 4th IEEE International Colloquium on Information Science and Technology (CiSt)* (IEEE, 2016), pp. 610–614

17. M.R. Anawar, S. Wang, M. Azam Zia, A.K. Jadoon, U. Akram, S. Raza, Fog computing: an overview of big iot data analytics. Wireless Commun. Mob. Comput. **2018** (2018)

18. S. Li, M.A. Maddah-Ali, A.S. Avestimehr, Coding for distributed fog computing. IEEE Commun. Magaz. **55**(4), 34 (2017)

19. M. Gupta, Fog computing pushing intelligence to the edge. Int. J. Sci. Technol. Eng **3**(8), 4246 (2017)

20. R. Mahmud, R. Kotagiri, R. Buyya, Fog computing: a taxonomy, survey and future directions, in *Internet of everything* (Springer, 2018), pp. 103–130

21. N. Kotilainen, M. Weber, M. Vapa, J. Vuori, Mobile chedar-a peer-to-peer middleware for mobile devices, in *Third IEEE International Conference on Pervasive Computing and Communications Workshops* (IEEE, 2005), pp. 86–90

22. W. Lee, K. Nam, H.G. Roh, S.H. Kim, A gateway based fog computing architecture for wireless sensors and actuator networks, in *2016 18th International Conference on Advanced Communication Technology (ICACT)* (IEEE, 2016), pp. 210–213

23. H. Shi, N. Chen, R. Deters, Combining mobile and fog computing: using coap to link mobile device clouds with fog computing, in *2015 IEEE International Conference on Data Science and Data Intensive Systems* (IEEE, 2015), pp. 564–571

24. D. Poola, M.A. Salehi, K. Ramamohanarao, R. Buyya, A taxonomy and survey of fault-tolerant workflow management systems in cloud and distributed computing environments, in *Software Architecture for Big Data and the Cloud* (Elsevier, 2017), pp. 285–320
25. B. Schmidt, L. Wang, D. Galar, Semantic framework for predictive maintenance in a cloud environment. Procedia CIRP **62**, 583 (2017)
26. K. Hong, D. Lillethun, U. Ramachandran, B. Ottenwälder, B. Koldehofe, Mobile fog: A programming model for large-scale applications on the internet of things, in *Proceedings of the Second ACM SIGCOMM Workshop on Mobile Cloud Computing* (2013), pp. 15–20
27. V. Cardellini, V. Grassi, F.L. Presti, M. Nardelli, On qosaware scheduling of data stream applications over fog computing infrastructures, in *2015 IEEE Symposium on Computers and Communication (ISCC)* (IEEE, 2015), pp. 271–276
28. C. Dsouza, G.J. Ahn, M. Taguinod, Policy-driven security management for fog computing: preliminary framework and a case study, in *Proceedings of the 2014 IEEE 15th International Conference on Information Reuse and Integration (IEEE IRI 2014)* (IEEE, 2014), pp. 16–23
29. H.M. Hashemian, State-of-the-art predictive maintenance techniques. IEEE Trans. Inst. Measur. **60**(1), 226 (2010)
30. T. Cerquitelli, D. Bowden, A. Marguglio, L. Morabito, C. Napione, S. Panicucci, N. Nikolakis, S. Makris, G. Coppo, S. Andolina, A. Macii, E. Macii, N. O'Mahony, P. Becker, S. Jung, A fog computing approach for predictive maintenance, in Advanced Information Systems Engineering Workshops - CAiSE, International Workshops, Rome, Italy, June 3–7, 2019, Proceedings, Lecture Notes in Business Information Processing, vol. 349, ed. by H.A. Proper, J. Stirna (Springer, 2019). Lecture Notes in Business Information Processing **349**, 139–147 (2019)
31. S. Panicucci, N. Nikolakis, T. Cerquitelli, F. Ventura, S. Proto, E. Macii, S. Makris, D. Bowden, P. Becker, N. O'Mahony, L. Morabito, C. Napione, A. Marguglio, G. Coppo, S. Andolina, A cloud-to-edge approach to support predictive analytics in robotics industry, Electronics **9**(3) (2020). https://doi.org/10.3390/electronics9030492. https://www.mdpi.com/2079-9292/9/3/492

Data-Driven Predictive Maintenance: A Methodology Primer

Tania Cerquitelli, Nikolaos Nikolakis, Lia Morra, Andrea Bellagarda,
Matteo Orlando, Riku Salokangas, Olli Saarela, Jani Hietala, Petri Kaarmila,
and Enrico Macii

Abstract Predictive maintenance aims at proactively assessing the current condition of assets and performing maintenance activities if and when needed to preserve them in the optimal operational condition. This in turn may lead to a reduction of unexpected breakdowns and production stoppages as well as maintenance costs, ultimately resulting in reduced production costs. Empowered by recent advances in the fields of information and communication technologies and artificial intelligence, this

T. Cerquitelli (✉) · L. Morra · A. Bellagarda · M. Orlando
Department of Control and Computer Engineering, Politecnico di Torino, Turin, Italy
e-mail: tania.cerquitelli@polito.it

L. Morra
e-mail: lia.morra@polito.it

A. Bellagarda
e-mail: andrea.bellagarda@polito.it

M. Orlando
e-mail: matteo.orlando@polito.it

N. Nikolakis
Laboratory for Manufacturing Systems & Automation, University of Patras, Patras, Greece
e-mail: nikolakis@lms.mech.upatras.gr

R. Salokangas · O. Saarela · J. Hietala · P. Kaarmila
VTT Technical Research Centre of Finland, Kemistintie 3, 02150 Espoo, Finland
e-mail: riku.salokangas@vtt.fi

O. Saarela
e-mail: olli.saarela@vtt.fi

J. Hietala
e-mail: jani.hietala@vtt.fi

P. Kaarmila
e-mail: petri.kaarmila@vtt.fi

E. Macii
Interuniversity Department of Regional and Urban Studies and Planning, Politecnico di Torino, Turin, Italy
e-mail: enrico.macii@polito.it

T. Cerquitelli et al. (eds.), *Predictive Maintenance in Smart Factories*,
Information Fusion and Data Science,
https://doi.org/10.1007/978-981-16-2940-2_3

chapter attempts to define the main operational blocks for predictive maintenance, building upon existing standards discusses and key data-driven methodologies for predictive maintenance. In addition, technical information related to potential data models for storing and communicating key information are provided, finally closing the chapter with different deployment strategies for predictive analytics as well as identifying open issues.

1 Introduction

Digitisation in production systems mandates a shift in existing business models. In the Industry 4.0 scenario, new technologies are incorporated in the modern manufacturing site, enabling a novel paradigm in which different systems are interconnected and synchronised to globally optimise and improve all aspects of the product system. This paradigm shift heavily relies on Information and Communication Technology (ICT) systems and Internet of Things (IoT) devices to produce, manage and store a large amount of information, that, however, cannot be easily exploited. Consequently, the manufacturing industry demands new methods and approaches in order to analyse those data that may help in maintenance activities [1, 2].

Given the high cost associated with maintenance activities, both in terms of equipment restoration and production downtime, preserving industrial equipment in working condition emerges as a key challenge for every production system [3]. In response, instead of carrying out predetermined maintenance activities that may or may not be required, condition monitoring approaches were introduced [4]. Such approaches allow for maintenance activities to be planned according to the actual operational status of a machine [5]. Maintenance activities can be thus be scheduled proactively to minimise costs and maximise production. Accordingly, conditions that can be harmful to the operational lifetime of any production equipment can be prevented and the ultimate ambition, that is, the deferral and optimisation of maintenance procedures, can be achieved.

Nevertheless, it is extremely challenging to identify the correct set of parameters or models that may be associated to potential failures, determine their critical values, and model them as part of the degradation process of the machine [6]. This challenge is only increased when considering the number and heterogeneity of equipment on a shop floor [7].

In this context, predictive data analytics techniques [8] are attracting increasing interest and are, without doubt, one of the key enabling technologies in the Industry 4.0 landscape. They leverage the availability of sensors and IoT devices that collect widespread and detailed data regarding the current status and operation of machines, robots, and plants. Data-driven models are widely applicable in this context, as they require in principle minimal knowledge of the underlying physical or chemical phenomenon, and can be built following a relatively standardised data science pipeline. They have shown high potential in predicting potential failures and optimise maintenance activities accordingly. Given the importance of a smooth and uninterrupted production process, predictive analytics examine large-scale data sets and find out correlations and patterns in order to predict future events and maintenance needs, like the Remaining Useful Life (RUL) of a machine tool equipment [9].

Exploiting Big Data generated by Industry 4.0 brings several challenges related to their management, processing, and organisation [10]. This chapter aims at presenting an overview of the key concepts related to predictive maintenance, with a specific focus on data-driven methodologies. Predictive maintenance is introduced within the framework provided by the European standard CEN-EN 13306. Subsequently, the key concepts related to data-driven methodologies (descriptive analytics and predictive analytics) are introduced. The advantages and disadvantages of data-driven vs. physics-based models are discussed.

The various phases of the construction of a data-driven pipeline are exemplified by the data-driven pipeline developed in the context of the SERENA project. This case study addresses the common scenario of a slowly degrading, multi-cycle industrial process and covers two of the most common tasks in predictive maintenance: fault prediction and RUL estimation.

The success of predictive maintenance does not rely solely on building effective data-driven models. It also requires an effective and scalable infrastructure in which such models can be integrated and deployed. Such infrastructure must handle critical functionalities such as the representation, storage, transfer and integration of measurement data. The MIMOSA Open Systems Architecture for Enterprise Application Integration (OSA-EAI), distributed by the non-profit MIMOSA trade organisation, is specifically designed to combine data from different sensors or systems, so that the right data goes to the right place for the right person. The chapter is therefore closed by discussing how the MIMOSA technology can be leveraged to build predictive maintenance systems.

The rest of this chapter is organised as follows: Sect. 2 refers to the building blocks for predictive maintenance and Sect. 3 describes key concepts of data-driven and physics-based modelling methodologies. Section 4 describes in detail the design of the SERENA data-driven pipeline. Sections 5 and 6 discuss in detail key aspects related to the deployment of predictive maintenance models, focusing on the MIMOSA architecture for enterprise application integration and on the SERENA micro-service architecture. Finally, the most promising research directions for next-generation predictive maintenance analytics are presented in Sect. 7.

2 Predictive Maintenance: Main Concepts and Key Standards

The purpose of maintenance standards is to help a company develop an effective maintenance strategy that avoids unforeseen downtime, but also seeks to consider the costs associated with the maintenance itself. Therefore, it is of paramount importance to first perform an FMECA analysis (IEC 60812) on the means of production or products to identify the most critical components for preventive maintenance. The company can then choose the maintenance strategy it deems appropriate based on CEN-EN 13306. Automation plays a big and significant role in today's maintenance. For this purpose, the ISO 13374 standard includes six steps that are intended to be automated, starting with data collection and ending with decision support. In addition, there are separate standards (ISO 17359, ISO 13381) related to diagnostics and prognostics, which are linking well to ISO 13374. The following sections deal in more detail with these maintenance-relevant standards.

2.1 The European Standard CEN-EN 13306

When adopting or implementing a maintenance strategy, it is crucial to understand what maintenance is, how it is defined, and what it concerns. An excellent starting point is provided by an established standard such as the European CEN-EN 13306, which defines maintenance as follows:

"Maintenance consists of all technical, administrative and management issues of the lifetime of the object designed to maintain or restore the object so that the object is capable of performing the required function [11]*."*

The overview of different maintenance strategies according to CEN-EN 13306 is presented in Fig. 1.

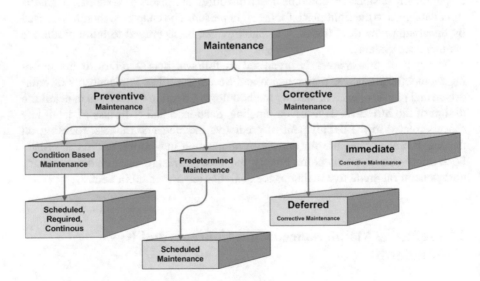

Fig. 1 Overview of maintenance strategies according to the CEN-EN 13306 standard

2.1.1 Preventive Maintenance

Preventive maintenance aims at reducing the probability of failures and degradation. It is divided into Condition-Based Maintenance (CBM) and Predetermined Maintenance, based on whether maintenance actions are carried out based on estimated machine state or at predetermined intervals.

In *Predetermined Maintenance*, machines are maintained at established intervals, measured in, e.g., elapsed calendar time, elapsed machine use hours, number of operational cycles, or number of kilometres travelled [12]. This maintenance strategy is suited for cases when degradation can be well predicted from such units of machine use, and when maintenance can be carried out in time, before degradation becomes too severe [13]. Predetermined maintenance places minimal requirements for instrumentation and data processing.

In *Condition-Based Maintenance* (CBM), maintenance schedules are established and updated based on the estimated current machine condition [12]. This maintenance strategy seeks to forecast future faults, failures, or problems in order to devise a proactive maintenance plan. Machine state can be evaluated and maintenance schedule updated either at scheduled intervals, on request, or continuously. Machine condition, and hence need for maintenance can be estimated with either data-driven or physics-based models, as will be explained in detail in Sect. 3.1.

Often the maintenance plan is optimised to minimise costs, e.g., by minimising downtime. CBM requires an initial investment on instrumentation and data processing, as well as operating costs. Hence it is also important to decide when a CBM strategy is cost-effective [12]. In a large system, CBM can be financially justified for some components or failure modes, predetermined maintenance for some other, and corrective maintenance for the remaining components or failure modes. CBM is recommended especially for machines that are critical for production or safety, or are capital-intense. Techniques that can be used to select the most appropriate maintenance approach, and whether CBM benefits offset its costs, are discussed in [13].

2.1.2 Corrective Maintenance

Corrective maintenance is a maintenance strategy in which corrective action is taken only when the fault actually occurs. The machine is then serviced to restore it to a condition where it can continue to carry out its required operation. This means that the machine or its component is not monitored proactively, but action is taken only when a fault condition occurs [12].

A corrective maintenance strategy is particularly well-suited for non-critical machines for which the cost of capital is low and the consequences of failure are not significant and there are no safety risks associated with this choice. In addition, it is important that the repair of the machine or its component is easy and quick [13].

In *deferred maintenance*, corrective actions are not performed immediately upon the occurrence of a failure, but are intentionally delayed so that the effects of maintenance on production can be minimised. This type of maintenance can be selected if the consequences of the fault do not significantly affect the operation of the machine [12].

Immediate maintenance is performed immediately after the occurrence of a fault in order to eliminate any significant production or safety consequences. This type of maintenance is applied in situations when the consequences of a fault are significant in terms of machine operation, safety or the environment [12], and where CBM is not applicable.

2.2 Guidelines and Requirements for Automated Maintenance Software

Information systems supporting diagnostics and prognostics are usually tailored for particular applications and largely incompatible so that they cannot exchange data with each other, let alone perform concerted data processing. In practice, this means that a significant amount of craftsmanship is required to implement comprehensive systems where information from many individual applications is mutually shared and deployed. The ISO 13374 standard [14] gives guidelines and requirements for software automating condition monitoring.

According to ISO 13374, the data flow of an automatic maintenance system consists of six functional blocks, which are illustrated in Fig. 2 [15]. ISO 13374 describes each of these blocks, including their functions and their interconnections. In this way, both the maintenance engineer and the operator are able to take advantage of compatible hardware and software.

Data-driven techniques may cover one or more of these blocks for extracting actionable knowledge on when and how to maintain a machine. Multiple different blocks (e.g., data manipulation, state detection, and prognostic assessment) may be merged within the same software.

2.3 Guidelines for Automated Condition Monitoring System and Diagnostics

The ISO 17359 standard [16] provides guidelines for automated condition monitoring and diagnostics, based on measured or calculated entities including, e.g., temperature, pressure, and vibration. Measurements and computed entities typically indicate machine condition in terms of availability, performance, and product quality. The ISO 17359 standard gives a list of factors to be considered when deciding a maintenance program for different machines, regardless of the type of function they perform. Therefore, it provides a systematic method for implementation of condition monitoring for a generic machine. The method aims at avoiding root causes

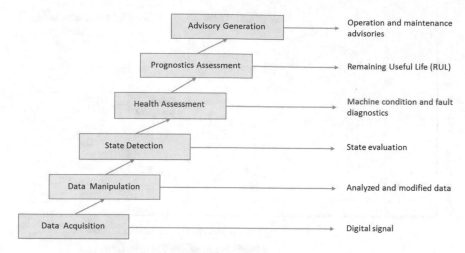

Fig. 2 Data processing blocks in standard ISO 13374 [14, 15]

of failures by providing instructions for maintenance, diagnostics, and prognostics. Improving the reliability of computed results is also considered.

The procedure described in ISO 17359 describes steps recommended for the implementation of CBM. The steps include, e.g. selection of appropriate maintenance strategy, selecting measurement locations, and selecting alarm criteria. The steps are presented as a flow chart with a number of feedback loops to improve operation based on experience gained. Cost-effect analysis is emphasised.

2.4 Guidelines for Prognosis

The ISO 13381 standard establishes a process for condition monitoring based on five steps: detection of deviation from normal conditions, diagnostics to discover the causes of faults, prognostics to predict their future development, decision support, and post-mortem analysis of any failure cases [17].

RUL is an estimate of the time remaining before the machine fails to perform acceptably; it is based on probabilistic prediction of failure modes either occurring or becoming too severe. A prognostic model is usually built on domain knowledge and recorded data of previous cases of fault progression. Figure 3 illustrates fault development and prediction of RUL [15, 17, 18].

As described in ISO 13381, establishing prognostics requires defining the failure or other condition whose occurrence is to be predicted, and defining the desired time horizon and level of confidence needed for decision-making. These definitions set requirements for a prognostic model to be developed to estimate the current machine state and to estimate RUL. A prognostic model can be either physics- or model-

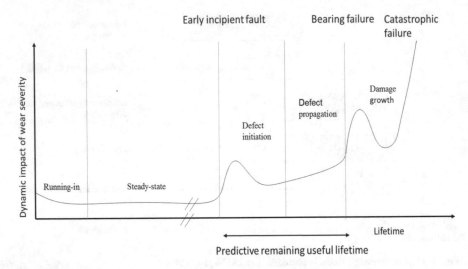

Fig. 3 Failure development and prediction of Remaining Useful Life [15, 17, 18]

based, purely data-driven, or a combination of both approaches, as will be detailed in Sect. 3.

3 Data-Driven Methodologies

This section introduces the principles behind data-driven predictive maintenance technologies, which are the main focus of this work. First, in Sect. 3.1 data-driven models will be compared against their physics-based counterparts, whereas Sect. 3.2 will provide a primer on data analytics techniques for predictive maintenance.

3.1 Data-Driven Modelling and Physics-Based Modelling

Prognostics and diagnostics models can be either based on the physics and chemistry of the operation and failure mechanisms of the asset, or data-driven, i.e., based on statistical or Machine Learning (ML) algorithms [19–21]. While the former are more interpretable, they also require a great domain expertise. Depending on the complexity of the machine, equipment or phenomenon under scrutiny, accurate physics-based models may be mathematically intractable or be computationally demanding. **On the contrary, data-driven models** are based on extracting the implicit knowledge available from past measurements, and tend to have wider applicability.

The respective advantages of data-driven (also known as empirical or "black box") and physics-based (also known as first principles or "white box") modelling

Fig. 4 Advantages of data-driven versus physics-based modelling

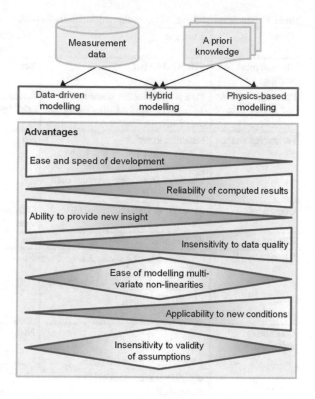

approaches are depicted in Fig. 4. Data-driven models are in general faster and cheaper to develop, and can consequently be developed for more applications given a fixed amount of resources. They are also able to provide new insights on relevant phenomena including, e.g., degrading factors that are present only in a few installations of similar equipment or structures.

Physics-based models, instead, are bound to our current understanding, and ability to mathematically model, the underlying phenomenon. However, they are generally more reliable, especially in extrapolation to conditions inadequately covered by training data, and are less sensitive to data quality issues and noise.

Multivariate and highly non-linear data-driven models may be unstable, e.g., small variations in the input cause a large variation in the output. This is especially true for highly non-linear, very large models such as deep neural networks [22]. On the other hand, physics-based models can be designed to enforce desirable stability properties. Physics-based models are also more directly interpretable, whereas the interpretability of data-driven models varies depending on the specific ML technique used.

Data analysis combining both approaches is called *hybrid* ("grey box") modelling [23–25]. Combining data analytics with domain knowledge has also been called the "third wave of AI" [26]. In hybrid modelling, knowledge of physics, chemistry,

Table 1 Examples of domain knowledge information that can be utilised to build hybrid models

Type	Example	Data-driven modelling	Hybrid modelling
Measurement readings	570 °C, 143 bar	Yes	Yes
Parameter values and design data	Pipe wall thickness when new = 15.6 mm. Manufacturing tolerance ± 0.3 mm	As measurement readings	Yes
Functional shapes	Reaction kinetics $dP/dt = k(T)A^n \times B^m$ with parameters $k(T)$, n, and m identified from measurement data P, T, A, and B	No	Yes
Constraints	$T_1 < T_2 < 100^o C$	Only in data validation	Yes
Equations	$F = m \times a$	No	Yes
Partial models		No	Yes
Qualitative information	• x causes y, not vice versa • y=f(x) is monotonously increasing • Measurement A is much more reliable than measurement B	No	Yes

etc., of the relevant phenomenon, machine or plant is utilised. The more the model relies on pre-existing domain knowledge, the less it depends on the quality and completeness of the available data. The optimal balance between these information sources is specific to the application characteristics and business constraints. The amount, type, and quality of the data available is also crucial to find the right balance between the two approaches. Figure 1 shows examples of which, and how, domain knowledge can be exploited when building hybrid models. Data-driven and physics-based elements can be combined in a hybrid model in a number of ways. Available approaches, which can be applied either individually or jointly, include:

- Modelling some parts of a system with a data-driven approach, and other parts with a first principles approach.
- Applying the first principles approach to reach some level of modelling accuracy, and then improving accuracy with data-driven techniques. For instance, a data-driven model can be identified from the residuals of a first principles model.

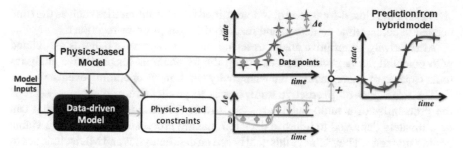

Fig. 5 Hybrid model, in which a data-driven model is used to refine the baseline prediction from a physics-based model. Figure adapted from [23]

- Defining constraints (such as monotonicity, functional shapes, or possible ranges of parameter values) for data-driven modelling based on a priori knowledge. This makes model identification computationally more demanding, as global optimisation techniques will be needed.

Figure 5 depicts a configuration where data-driven modelling is used to improve the accuracy of physics-based models. Since the physics-based model provides a robust baseline, care should be taken not to weaken this robustness through the data-driven model. As previously mentioned, some ML models are highly prone to instability, e.g., minor changes in the inputs have been observed to cause major changes in the model's output values [22]. This issue can be tackled by imposing physics-based constraints to the outputs of data-driven models.

3.2 Predictive and Descriptive Data Analytics

Data-driven approaches can be broadly classified in two categories: *descriptive analytics* and *predictive analytics*. Both are built upon elementary data manipulation activities which filter and map the collected data to higher level events and features, that are then fed into ML algorithms to build a statistical model of the phenomenon of interest.

Descriptive analytics encompasses a wide range of algorithms and tools that describe and summarise acquired data, and process them in order to uncover previously hidden, non-trivial and (possibly) causal relationships among events. In other words, descriptive analytics "reinterprets the past" to enable understanding of the manufacturing process, discover recurrent or unusual patterns, identify points of failure, and possibly trace them back to their underlying causes.

Descriptive analytics can detect an event after it has occurred, e.g., the presence of a fault on a piece of equipment. A corrective maintenance intervention can be triggered by the descriptive analytics technique. Corrective maintenance can be immediate or deferred as discussed in Sect. 2.1.2. Within this context, the responsiveness of the closed-loop control is determined by relevant metrics such as the time needed to detect (time-to-detect) and react to the fault (time-to-reaction).

Alternatively, descriptive analytics techniques can recognise patterns associated with potential *future* problems, thus laying the foundations for predictive analytics tools capable of understanding the when, why and how of upcoming events.

The main goal of **predictive analytics** is to predict future outcomes, such as the probability of a fault. Such services provide forecasting capabilities that can be ultimately deployed to improve decision making processes by suggesting viable coping strategies. They achieve this goal based on data analytics and ML techniques to extract aggregated forms of knowledge from raw data. As such, the potential business value is much higher for predictive rather than descriptive analytics, and the rest of this chapter will focus on their design and deployment. Predictive analytics techniques can be first categorised based on the specific goal or outcome they target, which in turns determines the class of ML techniques to be used and, more importantly, the cost function that will be optimised during their training.

For instance, many authors frame the predictive maintenance problem as predicting the *RUL* of the target asset [20, 21, 27–33]. RUL estimation is thus formulated as a *regression* problem where the output variable is a continuous (real or discrete) time value, typically defined as the time or number of production cycles left before the end of its expected operational life [27]. A more in-depth discussion of techniques for RUL estimation is given in Sects. 3.4 and 4.5.

Predictive maintenance can be also framed as a *Fault Prediction (FP)* problem [19, 34–39]. The goal here is to estimate the probability that a given asset will fail within a pre-determined time horizon (e.g., one day, week or month from the current time), which sets the problem as a *classification* task. FP models yield estimates with a coarser time granularity compared to RUL estimation, and for this reason are often easier to build and train. When framed in the context of descriptive analytics, the FP task becomes the Fault Detection (FD) task. Within the many emerging applications of FD, computer vision-based inspection techniques deserve a special mention: their goal is distinguish correctly manufactured or assembled components from faulty ones [40–42]. Both FP and FD tasks can be framed as either *binary* (i.e., presence or absence of fault) or *multi-class*, where different categories of faults are classified based either on their severity or their underlying causes [35].

3.3   Model Building, Deployment, and Updating

The three phases through which predictive analytics services are built and operated are *Model Building*, *Model Deployment* and *Model Updating*.

Model building. This phase consists in selecting and training a ML model based on historical data acquired ad-hoc or, preferably, during normal operating conditions. Model building consists of several phases including data preprocessing, feature engineering, feature selection, ML model selection, and finally training.

The predictive maintenance model can be built directly on features acquired from sensors, such as temperature or velocity measurements. In the case of complex, unstructured data, or in the presence of time series, it is usually preferrable to pre-process the raw data to extract more meaningful, higher-level characteristics or features. Feature engineering is a critical step in which the raw data is converted into informative features, usually relying on available domain knowledge. If the number of features is large (e.g., in the order of hundreds or thousands), a feature selection step may be necessary, also depending on the type of ML technique used. A practical case study is discussed in detail in Sect. 4.

After feature extraction and selection, the ML model is trained by fitting its param-eters to a training set based on the selected cost function to be optimised. For classi-fication tasks, such as FP/FD, a labeling phase may be necessary. The cost function depends on the type of task (regression vs. classification), but may also take into account operational constraints. For instance, different failures mode may be asso-ciated with different severity and costs, and therefore could be weighted differently.

A large number of ML algorithms are available and, in general, the best algorithm depends on the specific problem and dataset. Additionally, the complexity of the model must be carefully tuned to the problem at hand. Thus, the prediction perfor-mance crucially depends on a number of hyper-parameters whose value needs to be experimentally selected for the problem at hand. This tuning process can be long and time-consuming; however, as detailed in Sect. 4, **self-tuning strategies** (some-times denoted as **AutoML**) are being developed to reduce the manual effort, albeit at the expenses of a higher computational cost. It is important to evaluate model performance on a subset of the data that was not seen during training, to assess how well the model will perform in production and to select the model that generalises best to previously unseen data. For this purpose, the available dataset can be split into a training, validation and test set, using the validation set for model selection, and the test set for reporting the performance. Alternatively, one can resort to *K-fold cross-validation*, which splits the data into K folds and, for K times, alternatively uses a fold as test set and the other $K - 1$ as training set. Cross-validation is a viable alternative to a held-out test set when the data is scarce and expensive to acquire.

A wide variety of ML models can be used for predictive analytics, which are broadly categorised into *conventional ML* and *Deep Learning (DL)* algorithms. The term deep learning refers to a class of hyper-parametrised neural networks with a large number of layers (up to one hundred or more) stacked on top of each other. Their main advantage is that they are trained directly from raw data, with minimal preprocessing; features are not selected by the data scientist, but are rather learnt from data and jointly optimised along with the final classification/regression task. Deep neural networks are extremely effective for selected types of input data (e.g., for vision-based fault inspection), but are not always applicable or advantageous compared to other ML models. In particular, they require much larger training datasets and are more opaque than conventional ML techniques. In fact, selecting the proper trade-off between the accuracy of the predictions and their interpretability is a crucial factor in identifying the most appropriate ML model for a given use case.

Model building is a computationally intensive procedure, but with lenient require-ments in terms of latency. Hence, it is best performed *offline*, and offloaded to a data analytics platform with *batch processing* functionalities.

Model deployment. In this phase, the model is used to generate new predictions or *infer* new knowledge from data incoming from the distributed sensors. This requires performing online, real-time predictions on streaming data acquired, gathered, and processed through a dedicated analytics infrastructure. The time interval that occurs between data acquisition and the event prediction, as well as between the event prediction and the resulting preventive actions, defines the level of proactiveness of the predictive maintenance strategy.

Model update. Albeit fundamental, this is the least developed of the phases in analytics pipeline. A common assumption that underpins the majority of existing ML algorithms is that the data samples are independent and identically distributed (i.i.d.) random variables, with a stationary distribution. In other words, they assume that the data that is fed to the model at inference time have the same statistical properties of the training data, and that such distribution will be stable over time. In predictive maintenance scenarios, this assumption is not likely to hold true, as the nature of the data may change over time, e.g., due to equipment degradation, different operators, and so forth. Additionally, collecting data sets that are representative of all possible ground truth labels and/or acquisition conditions may be too expensive or technically unfeasible. For this reason, the ML model will likely be exposed to data with different distributions, or belonging to unseen class labels that were not available at training time. In this case, the model is likely to deviate from its expected performance and may lead to wrong outcomes [43].

Model updating exploits continuously acquired data from the field to partially retrain the model; it is needed to ensure that the performance model does not degrade over a sustained period of time [19, 21, 28, 36, 37, 43]. Both *passive* and *active* strate-gies have been proposed, depending on whether the update is periodic or triggered by some measure of model *degradation*. How to effectively detect model degradation in the field is still an active research area [43].

3.4   RUL Estimation: Key Concepts

The RUL of an asset is a widely used indicator to describe the conditions of an industrial asset or machine, and therefore a key data analytics task in the realm of predictive maintenance. It can defined as the time left to the end of the asset's oper-ational life, with respect to the functions for which it was constructed or purchased for.

Creating a general purpose model for RUL estimation, either physics-based or data-driven, is extremely challenging for a variety of reasons [44]. The same asset experiences different degradation levels under different operating conditions, results in varying RUL values in the same time horizon. Longer prediction horizons are desired for improving the optimisation of maintenance operations. Furthermore, in many industrial applications, there is plenty of data from normal operations, but few, if any, data from failures. In addition, there are often multiple different failure modes, wide spectrum of operating conditions, and numerous machine designs. Overload events, depicted in Fig. 6, are among the most important events in RUL estimation. Evaluating the impact of rare events is often best based on a combination of data analytics and knowledge of the failure mechanisms. Despite these challenges, RUL predictions remains a key component towards an effective predictive maintenance strategy. In this section, we present a review of the main approaches and refer the reader to Sect. 4.5 for a practical use case.

Model identification techniques rely on domain knowledge and an understanding of the physics of failure mechanisms to identify their occurrence. Specifically, a model is built that describes machine conditions, identifying model parameters from data, and then the model is extrapolated to estimate the time at which the condition deteriorates beyond a threshold. When new data is available, the model and the prediction are updated.

It must be stressed that the estimation of machine condition, especially its future development includes uncertainties due to, e.g., inaccurately known material characteristics, simplifications made in modelling, limited availability of measurement data, and unknown future use patterns. The longer the prediction horizon, the wider the confidence interval denoting estimation uncertainty, as depicted in Fig. 7. Estimation uncertainty should be considered when scheduling maintenance.

In order to detect specific, well-defined failure types, simple *data analysis* techniques can be deployed. Often these techniques require specific types of measurement instruments to be installed. Typical examples include:

- Analysis of vibration data: condition of rotating machinery is evaluated based on spectral peaks at characteristic frequencies [18].
- Oil analysis: wear in gears is modelled and predicted based on measured counts of metal particles in lubrication oil [46].

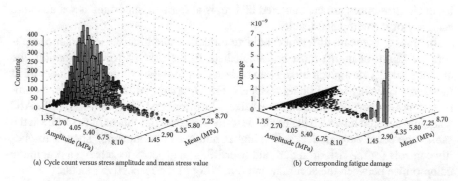

(a) Cycle count versus stress amplitude and mean stress value   (b) Corresponding fatigue damage

Fig. 6 Stress cycles of a concrete structure, and corresponding damage due to fatigue [45]. Cycles with high amplitudes are rare events, yet they are the most important ones when estimating RUL. Knowledge of fatigue physics is needed for reliable analytics

Fig. 7 The longer the
prediction horizon, the larger
the prediction uncertainty.
Development of machine
degradation is shown in fixed
blue line and its confidence
interval with dashed lines

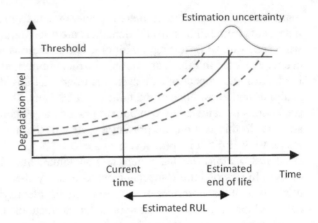

- Load and use profiling: damage due to fatigue in mechanical structures and welds
  is estimated by counting load cycles [47].

In many industrial cases, however, there is no obvious choice of model structure
to be deployed. Machine design and characteristics, their operating patterns, environmental factors, and characteristics of available data vary considerably. For this
reason, data-driven techniques are particularly appealing. A wide variety of statistical
models have been proposed, from linear regression to Hidden Markov Models and
deep neural networks. Several introductions and reviews of these techniques have
been published, covering both data-driven, physics-based, and hybrid approaches
[46, 48, 49]. However, most state-of-the-art algorithms for RUL estimation, such
as those based on Hidden Markov Models or Recurrent Neural Networks, require
extensive and complex modelling, as well as extensive training datasets, with limited efficiency. In addition, datasets available for training RUL estimation models
are rarely comprehensive enough for this approach, as data corresponding to failure
cases is rare. In many industrial cases, the robustness of the model, i.e., its reliability in circumstances not encountered before, is at least as important as the absolute
accuracy of the computed results.

In order to compare different candidate models, a quantitative performance measure (or figure of merit) must be established. In the case of RUL estimation, the
absolute quality of a given model can be assessed through common goodness of fit
measures, typically based on its residuals. For comparing different models, a common figure of merit is based on the Akaike's Information Criterion (AIC) [50]. AIC
takes into account how well the model describes the data (goodness of fit) while at the
same time penalizing models with a higher number of parameters. Thus, it implic-
itly controls for the risk of under- and overfitting, and seeks to achieve a pragmatic
compromise between the accuracy and reliability of the respective models.

Mathematically, AIC is based on information theory principles and estimates
the relative amount of information lost by a given model. This information loss is
evaluated in the sense of the Kullback–Leibler divergence, a measure used to estimate

how two probability distributions differ. Since the true process that generated the data is unknown, AIC estimates the relative quality of a model, related to other models, up to a dataset dependent offset. This means that AIC values are comparable between models trained and validated on the same dataset. In addition, AIC is able to establish the relative, but not the absolute quality of a model: in other words, the AIC will not detect cases where all models fit the training data poorly. Thus, it should be combined with statistical techniques that can quantify the performance of a model in absolute terms [50].

4 The SERENA Data-Driven Pipeline

In this section we present a scalable, data-driven pipeline for predictive maintenance developed in the context of the SERENA project. It addresses the common case study of a slowly degrading, multi-cycle industrial process, with the goal of predicting alarm or failure conditions. It exploits historical data to build a predictive analytics model.

The proposed data-driven pipeline is composed of several building blocks. The *data preparation* step (described in Sect. 4.1) refers to all strategies employed to describe and aggregate the cyclic time series data, both in the model *building* (training) and model *deployment* (inference) phases. The predictive model is selected and trained in the *semi-supervised data labelling* (described in Sect. 4.2) and the *predictive analytics* step (described in Sect. 4.3). Finally, during model deployment a *self-assessment* step is introduced to identify model degradation and trigger the *model update* process.

An important aspect when designing data-driven pipelines is how to reduce the cost of manual interventions by data scientists and domain experts, whose effort should most be mostly devoted to high-level, intellectual aspects such as designing the overall pipeline, setting objectives and requirements and interpreting results. Instead, the time devoted to repetitive tasks, such as data labelling and hyper-parameter optimisation, should be limited.

For this reason, the proposed pipeline features two important innovations. First, the effort devoted to data annotation is cut through the use of efficient semi-supervised techniques. Secondly, the proposed pipeline is able to self-tune the data-driven models to offload data scientists and domain experts from manual interventions. While the data preprocessing and aggregation parameters are rooted in the specific features of a given industrial process, and thus can be selected relatively easily by domain experts, the best algorithms and their hyper-parameters are automatically selected through an automatic grid search strategy. Self-tuning strategies can be applied to the feature selection, as well as to the actual ML model building phase.

4.1 Data Preparation

This component performs the common steps required to clean, transform, and aggregate the data collection under analysis. This step is essential to transform *raw measurement* into *Smart Data*, interpretable by users and domain experts, on which knowledge and insights can be built. To address the cyclic nature of the industrial process under exam, all data are aligned to fit a regular structure, in which each time series has a fixed time duration and represents one cycle. Padding is used to align all cycles to the same duration, i.e., the last value is repeated until the time slot is filled.

Data cleaning. This phase removes and detects outliers (i.e., extreme data points that differ significantly from other observations). Many ML models are sensitive to outliers, which may significantly degrade their performances at test time. However, a rigid, out-of-the-box, mathematical definition of outlier does not exist, and their identification ultimately depends on the knowledge of the underlying data generation process. In this case, outliers are selected based on the cycle length, removing those cycles that belong to the first and last decile of the distribution (in other words, the 10% shortest and 10% longest cycles are removed). This approach allows to remove cycles which were not true production cycles, e.g., test cycles, as confirmed by domain experts.

Feature extraction. Time-independent features are extracted from the raw time series recorded by the sensors by splitting them into contiguous chunks, which are then summarised by their statistical distribution (e.g., mean, sum, median, standard deviation, percentiles, skewness, kurtosis, number of elements over the mean, etc.). The split size is a parameter of the SERENA pipeline, which in our case was manually selected by the domain expert. Each chuck captures specific transient or steady states in the cyclic industrial process, whose duration can vary. This technique can be seamlessly applied to both fixed and variable split sizes, although a fixed chunk size is simpler to implement and works well in many practical scenarios.

Feature selection. This step allows to retain only the most useful features, while discarding those that are redundant or non-informative. In this pipeline, redundant features are removed based on two strategies: multicollinearity-based and correlation-based. The former removes features whose values can be trivially predicted by combining the values of other attributes in a multiple regression model. The latter instead calculates the mutual correlation of each feature pair and removes those features that have the highest average correlation, taking into account all the others.

Data aggregation. For many common slowly-degrading cyclic industrial processes, the alarm condition can manifest itself over a long period of time, in the order of hours or days. For this reason, the duration of a single cycle is too short with respect to the target degradation phenomena and its prediction horizon, which spans over many cycles, and there is little interest or benefit in predicting its alarming condition. This step aggregates the features calculated for each cycle over longer, multi-cycle time windows. To retain information about each individual state within the cycle, the aggregation is separately performed for each time chunk. SERENA captures changes in each feature by fitting a linear regression model over the aggregated multi-cycle period, and recording the resulting slope and intercept coefficients. Additionally, the minimum, maximum, average, and standard deviation are computed separately for each feature within the multi-cycle time windows. It is important to stress that both feature selection and aggregation preserve the meaning of each measurement. Thus, the overall SERENA pipeline is fully transparent, and its decisions accountable.

4.2 Semi-supervised Data Labelling

In order to build the predictive analytics model, it is necessary to label each production cycle according to the desired prediction class. This represents a widespread challenge in applying ML techniques to real-world settings, within the constraints of a limited supervision budget, and under the guidance of application-domain experts.

To tackle this challenge, SERENA relies on a semi-supervised approach to automatically partition the unlabelled dataset into a set of cohesive and well-separated groups (or clusters) of production cycles, each of which is assigned a different class label. Specifically, it leverages different *state-of-the-art clustering algorithms*, cohesively integrated via a *self-tuning strategy*.

Clustering. Clustering algorithms are employed to automatically discover a set of groups of production cycles characterised by similar properties. The self-tuning strategy not only automatically selects the optimal hyper-parameters for each algorithm, but also selects the one that finds the optimal partition, tailored to the specific data being analysed.

The proposed methodology integrates three partitional clustering algorithms: K-Means [51], Bisecting K-Means [52], and Gaussian Mixture Model (GMM) [53]. All clustering algorithms take as input a (user-defined) parameter k which represents the expected number of groups or partitions, but differ in their underlying mathematical models.

In the K-Means algorithm, each cluster is represented by its centroid, computed as the average of all samples in the cluster. Starting from an initial random configuration, it finds a partition that minimises the average inter-cluster distance (i.e., the distance between each sample in a cluster and the respective centroid). Although K-Means has a bias towards clusters of spherical shape, it obtains a reasonable solution in a limited amount of time in many real-life settings.

Bisecting K-Means does not search directly for an optimal global partition, but rather repeatedly focuses on portions of the dataset, which is then partitioned using K-means. The process is repeated until the desired number of groups is reached.

Finally, GMM is based on the assumption that each cluster can be modelled as a Gaussian distribution with unknown parameters (mean and variance). An iterative procedure is applied to estimate the parameters of each Gaussian distribution, as well as the sampling probability of each group, i.e., the probability that a given sample originates from each one of the k components.

Self-tuning strategy. A self-tuning strategy allows to select the best configuration with minimal or no manual intervention. It requires to establish an objective quality metric to compare different hyper-parameter settings and select the best performing one. A well-known quality metric for clustering algorithms is the Silhouette index [54], which measures how similar a given data point is to its own cluster (cohesion) compared to other clusters (separation). Here, a data point represents one production cycle.

The Silhouette index ranges from -1 to $+1$, where a positive (negative) value indicates the assigned cluster is a good (poor) match for a given data point, based on its distance from the cluster centroid, as well as from the neighboring clusters.

The goodness of a cluster is measured by the average Silhouette score of its data points. The ideal clustering algorithm partitions the data into a set of clusters with an average Silhouette equal to 1. In real-life settings with complex data distributions, an ideal clustering cannot be achieved, and Silhouette values in the range [0.2–0.3] are already considered acceptable values.

Labeling. Each cluster is manually labelled by a domain expert. To facilitate this task, SERENA characterises the statistical properties of each cluster by providing the *boxplot distribution of the top-10 intra-cycle features*. To identify the most relevant features, SERENA uses a Classification and Regression Tree (CART) [52] which is built using the same input features of the clustering algorithm and the cluster identifier as the target label. The first 10 features used as splits in the CART tree nodes are selected as the most relevant features. The boxplots provide visual support to domain experts to interpret the meaning of each cluster. Additionally, a few representative samples for each cluster are manually inspected. Their label assignments are then used for the remaining samples in each corresponding cluster.

4.3   Predictive Analytics

This building block consists of a classifier which assigns one of the labels discovered in Sect. 4.2 to new incoming data. As discussed in Sect. 3, this requires an offline *model building* phase in which the ML model is trained on historical data to extract the latent relations between the data and the prediction labels, representing different alarming conditions or failures.

Ensemble classification. SERENA is based on Gradient Boosted Tree Classifier [55] and Random Forest [56], two algorithms belonging to the family of ensemble classifiers. The best algorithm is selected based on a classification accuracy metric, such as the F-Score, calculated using the *Stratified K-Fold Cross Validation* and *Time Series Split Validation* techniques. The former is a standard cross validation approach where each fold retains the same proportion as the overall dataset distribution, hence the name stratified. The latter is based on selecting consecutive windows for training and testing; in practice, a training window with fixed origin and increasing width is selected at each iteration, and the corresponding test set is a fixed-size window right after the end of the training set. The algorithmic hyper-parameters are self-tuned through a grid search over a predefined parameter space.

4.4 Self Assessment of Model Drift

As discussed in Sect. 3.3, the performance of a predictive model can degrade over time. In industry and manufacturing, the distribution of collected data changes in time due to changes in machines or operators, equipment degradation, or differences in environmental factors. It is therefore of paramount importance to empower predictive analytics models with the ability to self-assess their performance and to trigger model retraining when necessary.

At the same time, in real use cases, it is impractical, expensive or downright unfeasible to collect new ground truth labels to evaluate the performance of a model in real-time. Hence, unsupervised techniques, that rely solely on the knowledge of the model inputs and outputs, must be employed.

The key concept behind SERENA self-assessment methodology is to identify variations in the geometrical distribution of the incoming data, with respect to historical data. The rationale behind this choice, confirmed by experimental observations, is that such changes corresponds to the presence of new data, belonging to an unknown class [43]. Given a dataset for which a set of class labels has been defined, it is possible to measure the cohesion and the separation of the data, and how it changes when new data becomes available. This can be achieved by combining traditional clustering metrics such as the Silhouette score, introduced in Sect. 4.2.

However, the calculation time needed to compute the traditional Silhouette score is not comparable with real-time analytics. For this reason, a new, scalable version called Descriptor Silhouette (DS) index was proposed to compute a fast, approximate estimation of the Silhouette score [57].

The degradation of the class c at the specific timestamp t is thus calculated as follows:

$$DEG(c, t) = \alpha \times \text{MAAPE}(DS_{t_0}, DS_t) \times \frac{N_c}{N} \qquad (1)$$

where the *Mean Arctangent Absolute Percentage Error* (MAAPE) [58] measures the discrepancy between the DS curve for the training data, and the same, possibly degraded curve calculated on the streaming data collected until the time t. The coefficient α can be equal to -1 or 1, depending on the relative values of DS_{t_0} and DS_t, and ensures that the degradation is positive. Further details are available in [43]. The MAAPE is weighted by the fraction of points predicted as belonging to class c (N_c) w.r.t. the total number of points (N).

4.5   RUL Estimation

Within the SERENA project we introduced a novel methodology for RUL estimation [59], which moves from the considerations and limitations of state-of-the-art models, discussed in Sect. 3.4. The approach consists in collecting data at different phases of an asset life cycle, building models to describe its behaviour at specified, reference time points, and identifying deviations from the normal or ideal operation profile. Data profiles collected at different stamps can be labelled to serve as reference for subsequent operations. For instance, data acquired at the initial asset deployment can be retained as the ideal operation profile, after which the operational behaviour begins to deteriorate.

Deviations from the nominal profile are based on a nominal set of key variables. In particular, the probability that each variable is sampled from the nominal distribution is estimated using the Gaussian kernel density estimation [60]. The degradation is assumed to be inversely proportional to the probability that each key variables is drawn from the reference nominal distribution observed during the ideal functioning of the machine: the lower the probability, the faster is asset degradation and thus, the lower is the final estimated RUL.

The set of key nominal variables can be preselected based on existing domain knowledge. Alternatively, it can be identified in a data-driven fashion by extracting those features that are most effective for pattern recognition and anomaly detection, in synergy with other predictive maintenance tasks. The degradation of the monitored equipment is quantified by measuring the drift between the nominal and observed key variable distributions, using the self-assessment technique described in Sect. 4.4. Once a drift is detected, the RUL indicator is correspondingly decreased.

In other words, SERENA estimates the RUL value by relying upon the joint *probability* distribution of the selected features with respect to their nominal or reference distribution. The *probability drift* is function of the time t that elapses from a reference time point. Mathematically, the RUL is defined as follows:

$$RUL(t) = \frac{1}{N_t} \sum_{x}^{X(t)} \left((100 - DEG(t)) \prod_{s}^{S} P(x_s \in K(X_s(t_0)) \right) \qquad (2)$$

where $X(t_0)$ refers to the nominal or ideal operational profile and $K(X_s(t_0))$ to the distribution of each key feature $s \in S$, estimated from the historical data $X(t_0)$. $X(t)$ represents the new signals collected from t_0 to time t, of cardinality N_t, and $P(x_s \in K(X_s(t_0)))$ is the probability that the feature S, estimated from the incoming data point x, is sampled from the nominal distribution $K(X_s(t_0))$. Finally, $DEG(t)$ is the overall degradation, which can be estimated based on the data collected from t_0 to t according to Equation (1).

In short, Equation (2) estimates the percentage of RUL, as a function of time, as the mean of the probabilities that each of the key nominal features is sampled from the ideal or nominal operational distribution, averaged over the new data points observed up until time t, and weighted by a factor that reflects the degradation of the asset at a given time. The higher the probability drift, the lower is the estimated RUL at time t.

The SERENA pipeline allows to estimate the RUL of any type of asset or machine using a customised, human-centered approach, which includes the knowledge of the domain expert in the identification of the key variables to be monitored. Since this experience is typically acquired during the production and operation of the equipment, it is not normally available or provided by the equipment manufacturer, unless the latter has the possibility to directly monitor the assets during their operational lifetime.

5 Data Formats and Models

When designing large-scale sensor networks for industrial use, attention needs to be given to the data formats and models used throughout the system to ensure its scalability and maintainability. The data formats should support a comprehensive enough context to connect each data point and event to a specific sensor or machine. The context for the collected data and sensors should be integrated with the data as close to the edge as possible. With sufficient context, machine problems or wear can be quickly pinpointed and circumvented or fixed by the maintenance crew. It also facilitates historical analysis, for example to train ML models to better detect potential faults.

5.1 A Linked Data Approach

JavaScript Object Notation (JSON) is a standard text-based human-readable information model for transmitting data between web applications. It is designed to be lightweight, language-independent, and easy and fast to parse and generate (Ecma International, 2017). A competitor for JSON is commonly considered to be the Extensible Markup Language (XML) [61]. However, XML is more verbose, slower to parse and does not work as well with JavaScript which is nowadays the most common programming language for web applications. For these reasons, JSON has become the most popular format in data exchange. A JSON data consists of attribute-value pairs. Each of the values can contain one of the predefined types: string, numeric, object, array, Boolean, and null. In the example seen in Fig. 8, attributes name and location contain string types, attribute time contains a numeric type, and attribute event contains an array of objects.

In a step towards semantic networking, JSON has since been expanded to the Linked Data (JSON-LD) standard (W3C, 2020). JSON-LD is a way to transfer machine interpretable data across a network with the introduction of Internationalised Resource Identifiers (IRI) [62]. The attributes within a JSON-LD document are mapped to a vocabulary with a context element and IRIs defined by it. With context and IRIs, machines can programmatically understand what kind of data and objects the document contains and act on them accordingly. Using JSON-LD also facilitates inter-system communication by giving an option to translate the internal vocabulary of a system to a common vocabulary. Other systems can then translate the data from the common vocabulary to their own internal data format. The common vocabulary also enables scaling up to larger scale systems and facilitates data sharing.

JSON-LD is designed to be fully compatible with any existing JSON parsers which trivializes extending to JSON-LD without breaking any existing non-compliant applications (W3C, 2020). However, only JSON-LD parsers can properly extract all information from the document. As can be seen in Fig. 9, the only difference to a standard JSON document (Fig. 8) is that some new attributes are added to give *context* to the data. Most importantly, the *context* attribute contains a link to the vocabulary used throughout the document. The vocabulary may contain instructions for specific attributes such as name, location or time, but not all attributes are necessarily covered in the vocabulary. In addition to the context attribute, *id* and *type* attributes are added to the example in Fig. 8. In JSON-LD a node is identified by its id attribute. This attribute could contain, for example, a unique ID number or an IRI. Additionally, a node is commonly given a *type* attribute to clarify what kind of object the other attributes are describing.

JSON-LD is a convenient way to transfer data from device to device or from system to system due to its capabilities in standardising the data structures under a single vocabulary. It also gives context to the transferred data and its origin, for example, linking sensors to the data produced by them. However, common existing vocabularies for JSON-LD documents do not exhaustively cover requirements for large-scale industrial applications. A vocabulary fitting for industrial sensor networks could be derived from a data model specifically built for storing all necessary information for operation and condition-based maintenance.

Fig. 8 Example of a JavaScript Object Notation (JSON) document

```
{
    "name": "punch_tool",
    "location": "hall",
    "events": [
        {
            "name": "startup",
            "time": 112602
        }
    ]
}
```

Fig. 9 JSON document from Fig. 8 extended with JSON-LD context

```
{
    "@context": "http://example.org/context.jsonld",
    "@id": "00234567",
    "@type": "Machine",
    "name": "punch_tool",
    "location": "hall",
    "events": [
        {
            "@type": "Event",
            "name": "startup",
            "time": 112602
        }
    ]
}
```

5.2 MIMOSA

In principle, condition-based maintenance requires a database system, as the data must be stored in a specific place so that it can be accessed later, knowing which machine the data in question applies to. It must also be possible to query the same position to make this information available, for example, for analytics. Operators, service personnel, logistics managers, OEMs, parts suppliers, and engineers aim to get condition information of the production equipment to the right people, in the right place, when needed. Unfortunately, the categorised information is typically distributed to different information models, on various platforms, and then further separated by data types. Typically, this style-separated information has to be collected from several different databases and requires an ad-hoc integration to combine these databases, which in addition use different data models. The MIMOSA Open Systems Architecture for Enterprise Application Integration (OSA-EAI) definitions combine the discrete data so that they can be found in the right place by the right person. The data model integrates design, maintenance, operation and reliability data to a single hierarchical structure. In the past, these sections have been completely separate from each other, and now it is finally possible to use only one data model and multiply the benefits of the information obtained when all data is in one place. The data can then be viewed synchronously and reports generated from it as discussed in [63–65].

MIMOSA is a non-profit trade organisation made up of providers and end users of industrial asset management systems. It develops and distributes open source MIMOSA OSA-EAI and MIMOSA OSA-CBM. The MIMOSA OSA-EAI is shown in Fig. 10, which clearly illustrates the benefits of its adoption. MIMOSA OSA-EAI defines data models and structures designed for registry, condition monitoring, reliability, maintenance, and work management, which are detailed in Table 2, as described by [15, 63].

MIMOSA OSA-CBM standardises data transfer in a condition-based maintenance system. It describes the six service blocks of proactive maintenance systems, Fig. 11) and the interfaces between them as defined in (ISO 13374-1 and ISO 13374-

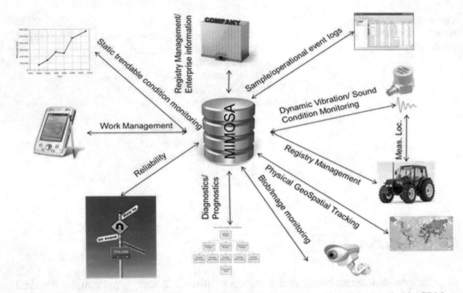

Fig. 10 MIMOSA Enterprise Application Integration connects functionalities needed in CBM

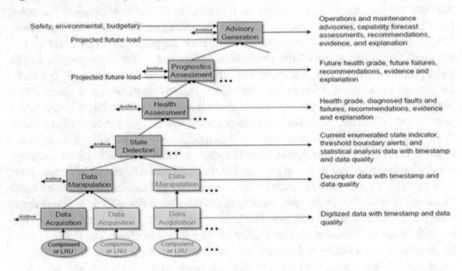

Fig. 11 MIMOSA Enterprise Application Integration connects functionalities needed in CBM

2). MIMOSA data structure is defined by CRIS (Common Relational Information Schema) relational database model. It has definitions for asset entities, their attributes with related types, and relations between entities. MIMOSA OSA-CBM specific blocks and ISO 13374 related functions are listed below. The blocks can request information from other functional blocks as needed. However, the data typically flows to adjacent blocks as described in ISO 13374-1 and [66, 67]. In particular the aforementioned blocks are as follows:

Table 2 MIMOSA database functionality (MIMOSA 2020)

Technology Types	Description
Registry Management (REG)	Definitions for enterprise, site, segment, asset, model and measurement location entities, and the hierarchical references between them. It is used by: • Original equipment manufacturer model information systems • Asset Registry Information Systems • Maintenance Management Systems
Reliability (REL)	Definition of hypothetical failure events for segments, assets, or serialisation assets. It is used by: • Reliability Information Systems • Failure mode, effects, and criticality analysis systems • Reliability-centred maintenance analysis systems
Work Management (WORK)	Definitions of work requests for segments or serialised assets, as well as work orders, steps, completion information, and pre-planned packages. It is used by: • Maintenance Management Systems
Diagnostics/ Prognostics/ Health Assessment (DIAG)	Definitions for diagnostics, prognostics, and health assessments. It is used by: • Diagnostic Systems • Prognostic Systems
Dynamic Vibration/ Sound Condition Monitoring (DYN)	Definitions for storing dynamic vibration and sound data used in condition monitoring. Also, it enables defining abnormal data alarm events. It is used by: • Vibration condition monitoring systems • Sound condition monitoring systems

- **Block 1—Data Acquisition**: Converts analogue physical quantities from sensors to digital format as defined in ISO 13374-1. The signal may be processed locally at the edge before being transferred to a central cloud database.
- **Block 2—Data Manipulation**: Performs signal analysis, computation of descriptor data, and derive virtual sensor readings from raw measurements, as defined in ISO 13374-1. Data is first validated and cleaned, and then used to compute information for condition monitoring of the asset.
- **Block 3—State Detection**: Detects the operational state (e.g., start-up, normal operation, shut-down) and searches for abnormalities as new data is acquired. It classifies whether the data is within a predetermined normality zone.

- **Block 4—Health Assessment**: Diagnoses faults from all state information and rate the current health of equipment or process.
- **Block 5—Prognostic Assessment**: Estimates future health states, predicts development of failure modes, and estimates RUL based on projected load profiles and current health assessment.
- **Block 6—Advisory Generation**: Generates operation and maintenance recommendations for optimizing the life and usage of the process and/or equipment. It utilizes health and prognostics assessment blocks.

6   Data Analytics Architecture

Deployment of predictive analytics services requires the orchestration of multiple building blocks. While Sect. 5 covers aspects related to data acquisition, manipulation, and exchange, the present section discusses the implementation of a data analytics architecture from a software engineering perspective. Starting from a description of the SERENA implementation, serving as an example of a modern, lightweight, and flexible data analytics architecture, further details are provided on software tools for analytics services (Sect. 6.1) and on the deployment of data analytics at the edge (Sect. 6.2).

The SERENA predictive maintenance analytics pipeline, described in Sect. 4, is built upon a lightweight and flexible micro-services architecture, where each application is implemented as an independent Docker service. Virtualisation tehcnologies like Docker create an agnostic middleware, insulating the service from the underlying host infrastructure, and ultimately improving scalability and resilience.

A standard practice for many non-IT industries is to rent cloud storage and servers to avoid capital and maintenance costs, while allowing fast scalability as needed. Needless to say, the monitored assets and the data-gathering sensors are distant from the cloud infrastructure. The SERENA architecture tackles this issue by wrapping edge applications as Docker services, in order to extend the IoT cloud concept to directly include distributed devices. Their deployment and lifecycle is managed as a Docker swarm, that is, a group of physical or virtual machines (workers or nodes) configured to join together in a cluster. The swarm is controlled by a swarm manager, responsible for distributing docker commands across the workers and handling their resources in an efficient way.

Wrapping edge functionalities as Docker services is a flexible approach to distribute application workloads in both effectively and efficiently. For instance, services can be transferred to the edge when needed to comply with network bandwidth or latency constraints, or can be executed on a centralised cloud infrastructure (either privately maintained, or remote public cloud), when needed to cope with high computational demands. Typically, data preprocessing and feature extraction is performed at the edge to avoid transferring the raw data, whereas the training of ML models is best executed centrally. The best trade-off depends on the type and size of the data, on the characteristics of the ML model, and service requirements in terms of latency and responsiveness.

6.1 Technologies and Tools for Analytics Services

Several software stacks and tools are available for the configuration, training, and deployment of both conventional and deep ML models, whose difference was briefly introduced in Sect. 3.3.

For conventional ML algorithms, the de-facto standard for implementing data-driven analytics on a distributed cloud platform is the Apache Spark's machine learning library MLlib [68], which includes distributed implementations of the most common algorithms introduced in Sect. 3.

Several open-source frameworks are available for implementing DL models, with TensorFlow [69] and PyTorch [70] being the current dominant frameworks. DL software stacks are key enablers for the efficient and effective design, training and deployment of deep neural networks. They implement automatic differentiation techniques to automatically calculate the gradients of millions of parameters during training, and transparently handle complex multi-dimensional matrix computations on high parallelism architectures such as Graphical Processing Units (GPUs), or the specialised Tensor Processing Units (TPUs). They also offer vast libraries of predesigned layers and network architectures, including pre-trained models for a variety of computer vision and natural language processing applications. Both frameworks support distributed training across multiple GPUs. Alternatively, a few community-driven projects, such as TFonSpark [71], enable the execution of TensorFlow or Pytorch models on Apache Spark.

6.2 Edge Analytics Applications

An edge analytics application executes a data analysis task in which at least part of data processing and analysis is performed close to where the data is collected. One of its benefits is the reduced latency in data analysis, enabling near real-time decision making. Furthermore, a larger amount of data can be collected and analysed because all the collected data does not have to be processed and centralised in a server. For these reasons, edge analytics also saves bandwidth as the need for data transfer is reduced, and prevents potentially sensitive data from being shared with third parties.

In applications which require high computational capacity, the use of edge analytics may be limited by the low computing power and insufficient random-access memory available at the edge. Nonetheless, given the advantages of performing analytics at the edge, several software tools and techniques are emerging.

Node-RED [72] is a popular edge computing solution which supports the execution of classical ML models and data transformation tasks. It provides many functionalities such as a visual interface, based on flow-based programming, to design data analytics pipelines, a large library of pre-designed ML algorithms, and standard interfaces to cloud servers.

Deploying DL analytics to the edge has also raised a lot of interest, especially for the analysis of images, video, and other high-dimensional signals, whose transferability is hindered by bandwidth, latency and privacy constraints. However, given the high computational costs associated to DL models, specialised architectures are required, both at the software and hardware level.

For this purpose, most diffuse DL frameworks offer lightweight, stripped-down versions of their inference engines, such as TensorFlow Lite and PyTorch Mobile, to execute DL models at the edge. These lightweight libraries do not support training, but are designed to reduce computational and memory overhead to a minimum to support single-core CPUs, micro-controllers and other devices with low computational budgets. Additionally, they include conversion tools to make ML models edge-compatible, reducing their memory and computational footprint.

It should be stressed, however, that bringing deep neural networks to the edge may require profound changes in the way a network is designed, trained or deployed. Typical examples are model partitioning and partial offloading, in which the execution of the model is shared between the edge nodes and the central cloud [73]. On the other hand, specialised hardware, such as the NVIDIA Jetson board, can be deployed to enhance the computational capabilities of edge nodes, although at the expenses of a higher cost [73]. Bringing DL to the edge is a highly researched, rapidly evolving topic with advances at the hardware, software and conceptual level.

In case there are no suitable edge analytics platforms available on the market, the organisation may need to develop its own platform, which can make edge analytics quite expensive. If there are only a few devices that should be analysed, the development costs are attributed only to a small number of installations. If, however, edge processing has to be scaled to a large number of installations, customised edge analytics can provide a cost-efficient solution [67].

As an example, we developed an edge device to monitor the health condition of bearings. It consists of low-cost components: a Raspberry Pi 3, a low-pass filter, an analogue to digital converter and an accelerometer. The constructed low-cost system measures the vibration acceleration of a bearing, processes the data sampled with a relatively high frequency, and sends the analysis results to a centralised server. The analytics of the health condition of the bearing is implemented using the VTT O&M Analytics Toolbox [74].

7   Discussion and Future Research Directions

In this chapter, the key concepts related to data-driven predictive maintenance methodologies are introduced and exemplified by means of the proposed SERENA pipeline for fault prediction and RUL estimation. Moving from those results, here, key emerging challenges and opportunities are discussed. These challenges must be addressed to enable value creation from data in the manufacturing industry and rip the most benefits from data-driven methodologies.

Interpretable prediction models. The most interesting predictive algorithms in the context of Industry 4.0 are those that yield interpretable models and predictions. Such models can bring novel knowledge and insights about the underlying industrial process and are easier to monitor during deployment, as malfunctionings can be more easily traced to their root causes. However, at the state-of-the-art some applications, especially those based on the interpretation of unstructured data such as images, text, and sound, can be solved with greater accuracy using DL techniques, briefly introduced in Sect. 3.3. Deep neural networks are highly non-linear, opaque, and often unstable models. Novel strategies that can shed light on the internal functioning of black-box algorithms are necessary to strike a better balance between accuracy and accountability in predictive models for Industry 4.0 applications. For instance, many techniques have been proposed to provide post-hoc explanations, e.g., by investigating the relationship between the input data and the model predictions.

Dealing with class imbalance. As anticipated in Sect. 3.4, data generated and collected in manufacturing and industrial contexts is usually highly skewed towards patterns associated with ordinary behaviour, a desirable property in practice as it describes a healthy operating system. On the other hand, data associated with failure modes is often scarce and may not represent all possible failures. This level of class imbalance becomes critical for the application of most machine learning techniques, which require many faulty samples to be able to detect patterns useful for their prediction. Training techniques that can help coping with data imbalance are needed to ensure that ML models operate correctly in, or are able to identify, unexpected events or situations, not represented in their training sets, while preventing them from converging to trivial solutions, such as predicting always the majority class.

Privacy-preserving data-driven methodology. Data-driven methodologies benefit from the collection of large datasets representative of a variety of different operating conditions. However, privacy and intellectual property issues associated with data collection and transfer are critical aspects that prevent the adoption and deployment of data-driven solutions in intelligent manufacturing. Federated learning is an emerging distributed machine learning approach which enables collaborative training on decentralised datasets [75]. In this paradigm, several sites can collaborate by training machine learning models locally with their own datasets, and then sharing the trained models to produce global models that can indirectly access a large corpus of data without requiring explicit data sharing. So far, this approach has shown comparable performance to traditional approaches while preserving data privacy. Nonetheless, several challenges must be overcome to effectively leverage such methodologies in Industry 4.0, and guarantee that the resulting predictive models achieve the same performance of a centralised trained model with access to all data.

Acknowledgements This research has been partially funded by the European project "SERENA – VerSatilE plug-and-play platform enabling REmote predictive mainteNAnce" (Grant Agreement: 767561).

References

1. K. Alexopoulos, S. Makris, V. Xanthakis, K. Sipsas, G. Chryssolouris, A concept for context-aware computing in manufacturing: the white goods case. Int. J. Comput. Integr. Manuf. **29**(8), 839 (2016). https://doi.org/10.1080/0951192X.2015.1130257
2. A. Sestino, M.I. Prete, L. Piper, G. Guido, Internet of Things and Big Data as enablers for business digitalization strategies. Technovation **98** (2020). https://doi.org/10.1016/j.technovation.2020.102173
3. A. Listou Ellefsen, E. Bjørlykhaug, V. Æsøy, S. Ushakov, H. Zhang, Remaining useful life predictions for turbofan engine degradation using semi-supervised deep architecture. Reliability Eng. Syst. Safety **183**, 240 (2019). https://doi.org/10.1016/j.ress.2018.11.027
4. K. Hendrickx, W. Meert, Y. Mollet, J. Gyselinck, B. Cornelis, K. Gryllias, J. Davis, A general anomaly detection framework for fleet-based condition monitoring of machines. Mech. Syst. Signal Process. **139** (2020). https://doi.org/10.1016/j.ymssp.2019.106585
5. K. Mykoniatis, A real-time condition monitoring and maintenance management system for low voltage industrial motors using internet-of-things. Proc. Manuf. **42**, 450 (2020). https://doi.org/10.1016/j.promfg.2020.02.050
6. T. Mohanraj, S. Shankar, R. Rajasekar, N.R. Sakthivel, A. Pramanik, Tool condition monitoring techniques in milling process-a review. J. Mater. Res. Technol. **9**(1), 1032 (2020). https://doi.org/10.1016/j.jmrt.2019.10.031
7. Agile shopfloor organization design for industry 4.0 manufacturing. Proc. Manuf. **39**, 756 (2019). https://doi.org/10.1016/j.promfg.2020.01.434
8. T. Cerquitelli, D.J. Pagliari, A. Calimera, L. Bottaccioli, E. Patti, A. Acquaviva, M. Poncino, Manufacturing as a data-driven practice: methodologies, technologies, and tools. Proc. IEEE **109**(4), 399 (2021). https://doi.org/10.1109/JPROC.2021.3056006
9. M.C.M. Oo, T. Thein, An efficient predictive analytics system for high dimensional big data. J. King Saud Univ.-Comput. Information Sci. (2019). https://doi.org/10.1016/j.jksuci.2019.09.001
10. L. He, M. Xue, B. Gu, Internet-of-Things enabled supply chain planning and coordination with big data services: certain theoretic implications. J. Manage. Sci. Eng. **5**(1), 1 (2020). https://doi.org/10.1016/j.jmse.2020.03.002
11. N. EN, 13306, Terminologie de la maintenance (2010)
12. M. Bengtsson, *Condition Based Maintenance Systems: An Investigation of Technical Constituents and Organizational Aspects* (Citeseer, 2004)
13. A. Starr, A Structured Approach to the Selection of Condition Based Maintenance. In *Fifth International Conference on FACTORY 2000 - The Technology Exploitation Process* (IET 1997), pp. 131–138. https://doi.org/10.1049/cp:19970134
14. B. ISO. 13374-1–condition monitoring and diagnostics of machines data processing, communication and presentation part 1: General guidelines, engineering 360
15. R. Salokangas, E. Jantunen, M. Larrañaga, P. Kaarmila, MIMOSA strong medicine for maintenance, in *Advances in Asset Management and Condition Monitoring* (Springer, 2020), pp. 35–47
16. I. ISO, *17359: 2011-Condition monitoring and diagnostics of machines-general guidelines* (Tech. rep, Technical report, 2011)
17. ISO, ISO 13381-1: 2015: Condition monitoring and diagnostics of machines–prognostics–part 1: general guidelines (2015)
18. I. El-Thalji, E. Jantunen, Fault analysis of the wear fault development in rolling bearings. Eng. Failure Anal. **57**, 470 (2015). https://doi.org/10.1016/j.engfailanal.2015.08.013
19. M. Canizo, E. Onieva, A. Conde, S. Charramendieta, S. Trujillo, Real-time predictive maintenance for wind turbines using Big Data frameworks, in *2017 IEEE International Conference on Prognostics and Health Management (ICPHM)* (2017), pp. 70–77. https://doi.org/10.1109/ICPHM.2017.7998308
20. Z. Liu, N. Meyendorf, N. Mrad, The role of data fusion in predictive maintenance using digital twin, in *AIP Conference Proceedings* 1949 (April 2018). https://doi.org/10.1063/1.5031520

21. M. Compare, P. Baraldi, E. Zio, Challenges to IoT-enabled predictive maintenance for industry 4.0. IEEE Internet of Things J., 1 (2019). https://doi.org/10.1109/JIOT.2019.2957029

22. C. Szegedy, W. Zaremba, I. Sutskever, J. Bruna, D. Erhan, I. Goodfellow, R. Fergus, Intriguing Properties of Neural Networks. In *2nd International Conference on Learning Representations* (ICLR) (2014)

23. M. Von Stosch, R. Oliveira, J. Peres, S.F. de Azevedo, Hybrid semi-parametric modeling in process systems engineering: past, present and future. Comput. Chem. Eng. **60**, 86 (2014). https://doi.org/10.1016/j.compchemeng.2013.08.008

24. U. Leturiondo, O. Salgado, L. Ciani, D. Galar, M. Catelani, Architecture for hybrid modelling and its application to diagnosis and prognosis with missing data. Measurement **108**, 152 (2017). https://doi.org/10.1016/j.measurement.2017.02.003

25. C.M. Chew, M. Aroua, M. Hussain, A practical hybrid modelling approach for the prediction of potential fouling parameters in ultrafiltration membrane water treatment plant. J. Industrial Eng. Chem. **45**, 145 (2017). https://doi.org/10.1016/j.jiec.2016.09.017

26. J. Launchbury, A DARPA perspective on artificial intelligence, in *DARPA Slides* (2017)

27. S. Goto, K. Tsukamoto, On-line residual life prediction including outlier elimination for condition based maintenance. Int. J. Innov. Comput. Information Control **8**(3B), 2193 (2012)

28. G.A. Susto, J. Wan, S. Pampuri, M. Zanon, A.B. Johnston, P.G. O'Hara, S. McLoone, An adaptive machine learning decision system for flexible predictive maintenance, in *2014 IEEE International Conference on Automation Science and Engineering (CASE)* (2014), pp. 806–811. https://doi.org/10.1109/CoASE.2014.6899418

29. G.A. Susto, A. Beghi, Dealing with time-series data in predictive maintenance problems, in *2016 IEEE 21st International Conference on Emerging Technologies and Factory Automation (ETFA)* (2016), pp. 1–4. https://doi.org/10.1109/ETFA.2016.7733659

30. A. Kanawaday, A. Sane, Machine learning for predictive maintenance of industrial machines using IoT sensor data, in *2017 8th IEEE International Conference on Software Engineering and Service Science (ICSESS)* (2017), pp. 87–90. https://doi.org/10.1109/ICSESS.2017.8342870

31. J. Yan, Y. Meng, L. Lu, L. Li, Industrial big data in an industry 4.0 environment: challenges, schemes, and applications for predictive maintenance. IEEE Access **5**, 23484 (2017). https://doi.org/10.1109/ACCESS.2017.2765544

32. F.D.S. Lima, F.L.F. Pereira, L.G.M. Leite, J.P.P. Gomes, J.C. Machado, Remaining useful life estimation of hard disk drives based on deep neural networks, in *2018 International Joint Conference on Neural Networks (IJCNN)* (2018), pp. 1–7. https://doi.org/10.1109/IJCNN.2018.8489120

33. P. Anantharaman, M. Qiao, D. Jadav, Large scale predictive analytics for hard disk remaining useful life estimation, in *2018 IEEE International Congress on Big Data (BigData Congress)* (2018), pp. 251–254. https://doi.org/10.1109/BigDataCongress.2018.00044

34. B. Kroll, D. Schaffranek, S. Schriegel, O. Niggemann, System modeling based on machine learning for anomaly detection and predictive maintenance in industrial plants, in *Proceedings of the 2014 IEEE Emerging Technology and Factory Automation (ETFA)* (2014), pp. 1–7. https://doi.org/10.1109/ETFA.2014.7005202

35. G.A. Susto, A. Schirru, S. Pampuri, S. McLoone, A. Beghi, Machine learning for predictive maintenance: a multiple classifier approach. IEEE Trans. Industrial Informatics **11**(3), 812 (2015). https://doi.org/10.1109/TII.2014.2349359

36. J. Xiao, Z. Xiong, S. Wu, Y. Yi, H. Jin, K. Hu, Disk failure prediction in data centers via online learning, in *Proceedings of the 47th International Conference on Parallel Processing* (Association for Computing Machinery, New York, NY, USA, 2018), ICPP 2018. https://doi.org/10.1145/3225058.3225106

37. D. Apiletti, C. Barberis, T. Cerquitelli, A. Macii, E. Macii, M. Poncino, F. Ventura, iSTEP, an integrated self-tuning engine for predictive maintenance in industry 4.0, in *IEEE ISPA/IUCC/BDCloud/SocialCom/SustainCom 2018, Melbourne, Australia, December 11–13, 2018* (2018), pp. 924–931. https://doi.org/10.1109/BDCloud.2018.00136

38. N. Amruthnath, T. Gupta, A research study on unsupervised machine learning algorithms for early fault detection in predictive maintenance, in *2018 5th International Conference on*

Industrial Engineering and Applications (ICIEA) (2018), pp. 355–361. https://doi.org/10.1109/IEA.2018.8387124

39. M. Fernandes, A. Canito, V. Bolón-Canedo, L. Conceição, I. Praça, G. Marreiros, Data analysis and feature selection for predictive maintenance: a case-study in the metallurgic industry. Int. J. Information Manage. **46**, 252 (2019). https://doi.org/10.1016/j.ijinfomgt.2018.10.006

40. J. Masci, U. Meier, D. Ciresan, J. Schmidhuber, G. Fricout, Steel defect classification with Max-Pooling Convolutional Neural Networks, in *The 2012 International Joint Conference on Neural Networks (IJCNN)* (2012), pp. 1–6. https://doi.org/10.1109/IJCNN.2012.6252468

41. D. Weimer, B. Scholz-Reiter, M. Shpitalni, Design of deep convolutional neural network architectures for automated feature extraction in industrial inspection. CIRP Annals **65**(1), 417 (2016). https://doi.org/10.1016/J.CIRP.2016.04.072

42. B. Staar, M. Lütjen, M. Freitag, Anomaly detection with convolutional neural networks for industrial surface inspection. Proc. CIRP **79**, 484 (2019). https://doi.org/10.1016/J.PROCIR.2019.02.123

43. T. Cerquitelli, S. Proto, F. Ventura, D. Apiletti, E. Baralis, Towards a real-time unsupervised estimation of predictive model degradation, in *Proceedings of the International Workshop on Real-Time Business Intelligence and Analytics, BIRTE 2019, Los Angeles, CA, USA, August 26, 2019* (2019), pp. 5:1–5:6. https://doi.org/10.1145/3350489.3350494

44. D.G. Kleinbaum, M. Klein, *Survival Analysis*, vol. 3 (Springer, 2010)

45. H. Ding, Q. Zhu, P. Zhang, Fatigue damage assessment for concrete structures using a frequency-domain method. Math. Problems Eng. (2014). https://doi.org/10.1155/2014/407193

46. M. Albano, E. Jantunen, G. Papa, *The MANTIS Book: Cyber Physical System Based Proactive Collaborative Maintenance* (River Publishers, 2019)

47. A.C.E. On Fatigue, Fracture, *Standard Practices for Cycle Counting in Fatigue Analysis* (ASTM International, 2005)

48. A.K. Jardine, D. Lin, D. Banjevic, A review on machinery diagnostics and prognostics implementing condition-based maintenance. Mech. Syst. Signal Process. **20**(7), 1483 (2006). https://doi.org/10.1016/j.ymssp.2005.09.012

49. N.H. Kim, D. An, J.H. Choi, *Prognostics and Health Management of Engineering Systems: An Introduction* (Springer, 2016)

50. D. Anderson, K. Burnham, *Model Selection and Multi-model Inference*, vol. 63 (2nd edn, Springer, NY) (2020), 10 (2004)

51. J.A. Hartigan, M.A. Wong, Algorithm as 136: A k-means clustering algorithm, J. R. Statistical Soc. Ser. C (Applied Statistics) **28**(1), 100 (1979)

52. P.N. Tan, M. Steinbach, V. Kumar, *Introduction to Data Mining*, 1st edn. (Addison-Wesley Longman Publishing Co., Inc, Boston, MA, USA, 2005)

53. B.G. Lindsay, Mixture models: theory, geometry and applications, in *NSF-CBMS Regional Conference Series in Probability and Statistics*, vol. 5, i (1995). http://www.jstor.org/stable/4153184

54. P.J. Rousseeuw, Silhouettes: A graphical aid to the interpretation and validation of cluster analysis. J. Comput. Appl. Math. **20**, 53 (1987)

55. J.H. Friedman, Greedy function approximation: a gradient boosting machine. Ann. Stat. **29**, 1189 (2000)

56. L. Breiman, Random forests. Mach. Learn. **45**(1) (2001)

57. F. Ventura, S. Proto, D. Apiletti, T. Cerquitelli, S. Panicucci, E. Baralis, E. Macii, A. Macii, A new unsupervised predictive-model self-assessment approach that scales, in *2019 IEEE International Congress on Big Data (BigData Congress)* (IEEE, 2019), pp. 144–148. https://doi.org/10.1109/BigDataCongress.2019.00033

58. S. Kim, H. Kim, A new metric of absolute percentage error for intermittent demand forecasts. Int. J. Forecasting **32**(3), 669 (2016)

59. S. Panicucci, N. Nikolakis, T. Cerquitelli, F. Ventura, S. Proto, E. Macii, S. Makris, D. Bowden, P. Becker, N. O'Mahony, L. Morabito, C. Napione, A. Marguglio, G. Coppo, S. Andolina, A cloud-to-edge approach to support predictive analytics in robotics industry. Electronics **9**(3) (2020). https://doi.org/10.3390/electronics9030492. https://www.mdpi.com/2079-9292/9/3/492

60. E. Parzen, On estimation of a probability density function and mode. Ann. Math. Statist. **33**(3), 1065 (1962). https://doi.org/10.1214/aoms/1177704472
61. W.W.W. Consortium, et al., *Extensive Markup Language* (xml) (2016)
62. W.W.W. Consortium, et al., *JSON-LD 1.0: A JSON-based Serialization for Linked Data* (2014)
63. MIMOSA 2020. www.mimosa.org
64. C. Hegedűs, P. Dominguez Arroyo, G. Di Orio, J. Luis Flores, K. Intxausti, E. Jantunen, F. Larrinaga, P. Malo, I. Moldován, S. Schneickert, *The MANTIS Reference Architecture* (River Publishers, Denmark, 2019), vol. 233, pp. 1361–1375
65. A.C. Márquez, V.G.P. Díaz, J.F.G. Fernández, *Advanced Maintenance Modelling for Asset Management* (2018)
66. M. Lebold, K. Reichard, C.S. Byington, R. Orsagh, Osa-cbm architecture development with emphasis on xml implementations, in *Maintenance and Reliability Conference (MARCON)* (2002), pp. 6–8
67. R. Salokangas, M. Larrañaga, P. Kaarmila, O. Saarela, E. Jantunen, MIMOSA for condition-based maintenance. Int. J. Condition Monit. Diagnostic Eng. Manage. p. to appear
68. A. Spark, *The Apache Spark Scalable Machine Learning Library*. Available: https://spark.apache.org/mllib/. Last access on May 2018 (2018)
69. Tensorflow. https://www.tensorflow.org/. Accessed 2020-04-23
70. Pytorch. https://pytorch.org/. Accessed 2020-04-23
71. Tensorflow on Spark. https://github.com/yahoo/TensorFlowOnSpark. Accessed 2020-04-23
72. Node-RED. https://nodered.org/. Accessed: 2020-04-20
73. J. Chen, X. Ran, Deep learning with edge computing: a review. Proc. IEEE **107**(8), 1655 (2019)
74. M. Larrañaga, R. Salokangas, P. Kaarmila, O. Saarela, Low-cost solutions for maintenance with a Raspberry Pi, in *Proceedings of the 30th European Safety and Reliability Conference and the 15th Probabilistic Safety Assessment and Management Conference* (Italy, Venice, 2020), pp. 1–6
75. Q. Yang, Y. Liu, T. Chen, Y. Tong, Federated machine learning: Concept and applications, ACM Trans. Intell. Syst. Technol. **10**(2), 12:1 (2019). https://doi.org/10.1145/3298981

Services to Facilitate Predictive Maintenance in Industry4.0

Nikolaos Nikolakis, Tania Cerquitelli, Georgios Siaterlis, Andrea Bellagarda, Enrico Macii, Guido Coppo, Salvatore Andolina, Florian Bartholomauss, Martin Plutz, and Kosmas Alexopoulos

Abstract The advent of Industry4.0 in automation and data exchange leads to a constant evolution towards intelligent environments, including an intensive adoption of Cyber-Physical System features. Thus, full integration of manufacturing IT and control systems with physical objects embedded with electronics, software, and sensors are being deployed. This new industrial model leads to a pervasive integration

N. Nikolakis (✉) · G. Siaterlis · K. Alexopoulos
Laboratory for Manufacturing Systems & Automation, University of Patras, Patras, Greece
e-mail: nikolakis@lms.mech.upatras.gr

G. Siaterlis
e-mail: siaterlis@lms.mech.upatras.gr

K. Alexopoulos
e-mail: alexokos@lms.mech.upatras.gr

T. Cerquitelli · A. Bellagarda
Department of Control and Computer Engineering, Politecnico di Torino, Turin, Italy
e-mail: tania.cerquitelli@polito.it

A. Bellagarda
e-mail: andrea.bellagarda@polito.it

E. Macii
Interuniversity Department of Regional and Urban Studies and Planning, Politecnico di Torino, Turin, Italy
e-mail: enrico.macii@polito.it

G. Coppo · S. Andolina
SynArea Consultants s.r.l., Turin, Italy
e-mail: guido.coppo@synarea.com

S. Andolina
e-mail: salvatore.andolina@synarea.com

F. Bartholomauss · M. Plutz
oculavis GmbH, Aachen, Germany
e-mail: bartholomaeus@oculavis.de

M. Plutz
e-mail: plutz@oculavis.de

© Springer Nature Singapore Pte Ltd. 2021
T. Cerquitelli et al. (eds.), *Predictive Maintenance in Smart Factories*,
Information Fusion and Data Science,
https://doi.org/10.1007/978-981-16-2940-2_4

of information and communication technology into production assets, generating massive amounts of data. Several software applications use such information for addressing versatile production requirements, such as predictive analytics, visualisation, and scheduling. Such services aim to facilitate daily maintenance activities and provide insight upon production assets and processes in a human-understandable way. In this chapter, the main services of the SERENA system exposed through its platform and towards supporting maintenance activities on the shopfloor are presented.

1 Introduction

Condition-based maintenance (CBM) can facilitate predictive maintenance policies as maintenance costs increase production efficiency. Through CBM approaches, the condition of a production asset is evaluated on real-world data. Based on the acquired information, its operational status is assessed, and its degradation estimated through suitable indicators, such as the Remaining Useful Life (RUL). The purpose is to identify imminent faults and proactively deal with them to preserve an asset in operational condition along with the overall production. To improve the efficiency of maintenance activities, appropriate planning is required, considering (i) the production activities and (ii) the real-life conditions. For example, many companies keep relying upon paper-based documentation for instructing operators on how to perform maintenance tasks. However, such approaches like the aforementioned lack flexibility that may have an impact on production efficiency and adaptability. Nevertheless, the integration of Industry 4.0 smart features can better support the daily activities of personnel and increase the efficiency of production systems. Hence, the SERENA platform and its services have been designed, developed, tested, and deployed in different and orthogonal industrial contexts, to address many of the industrial requirements in the context of Industry 4.0.

The SERENA platform allows collecting a large volume of data from industrial environments, smartly feeding analytics services to added-value applications, and support maintenance personnel through various augmented reality devices. Maintenance instructions and helpful information are visualised in multiple formats and through multiple interfaces depending on the use and operator type.

This chapter begins with a short review on software services related to predictive maintenance which is presented in Sect. 2. Then, Sect. 3 continues with the services implemented within the SERENA project in the context discussed in the previous section, tackling physics-based and data driven predictive analytics, scheduling and operator support functionalities. All services are part of the overall integrated SERENA system, aiming to support companies on their daily maintenance burdens and make them aware of the potential benefits stemming from predictive maintenance solutions.

2 Literature Review

Regarding predictive analysis, many studies have been carried out to develop efficient data management systems, data analytics engines, business processes, and risk assessment in Industry 4.0 [1]. Predictive analytics is based upon artificial intelligence techniques to design and implement predictive models to anticipate failures and promptly react to them. As a result, cost minimization and productivity improvement are achieved. Predictive maintenance techniques can reveal underlying information by analyzing historical data and predict the RUL of equipment and machines [2, 3]. The authors in [4] proposed PREMISES to forecast alarms regarding the maintenance of a machine or a part of it. Furthermore, to scale and improve data drift management over time, an unsupervised approach to detect and evaluate the degradation of classification and prediction models is discussed in[5]. The latter relies on a scalable variant of the Silhouette index, named Descriptor Silhouette, to quickly discover data drifting and the predictive model's weaknessess. The authors in [6] presented a flexible and scalable approach, named ISTEP (an integrated Self-Tuning Engine for Predictive maintenance). ISTEP relies on Big Data frameworks and data-driven methodologies that targeted the Industry 4.0 requirements to streamline the complexity of data-driven services' effective exploitation. However, various limitations have been identified in the approaches mentioned above. In particular, since the data reliability affects the result, poor quality data with a lot of noise and fault measurements significantly decrease each prediction's efficiency.

The estimation-prediction of the RUL is usually designed jointly with the scheduling application. The latter uses the prediction results to schedule the maintenance operations in line with the existing scheduling. The research activity presented in [7] describes a decision-making framework for selecting, according to the production plan, the appropriate time slot for assigning a maintenance activity to a relevant resource. The authors in [8] provides a complete survey on intelligent multi-criteria decision-making framework by evaluating alternative solutions for the shared human resources task scheduling.

A parallel research effort has been devoted to continuous condition monitoring of the production in the manufacturing. All information regarding the production line needs the appropriate visualization to be provided to the human [9]. The visualization techniques are focused on the Augmented Reality (AR), where the virtual information is generated to the user into the actual 'sensed' environment via specific hardware devices, [10], [11] as well as via Virtual Reality (VR), where the Industrial environment is simulated. All needed information is provided to the technicians. The report contains as many indicators correlated with the machine's status as instructions regarding the maintenance and the proper operation of each device [12]. According to Industry 4.0, visualization applications should provide personalized information to humans. The following methods use each user's characteristics and, along with artificial intelligence techniques, provide supporting instructions according to the users' preferences to support and motivate them to enhance their productivity [13, 14].

3 Predictive Maintenance Services in SERENA

The analytics services are intended to seamlessly support maintenance engineers and personnel by providing various and complementary functionalities. Data-driven and hybrid models for predictive maintenance are used as input measurement data and metadata information of the monitored machines. The two analytics services provide the primary input for the scheduling tool for planning the predictive maintenance activity concerning the analytics's service estimation. These functionalities' outcome is to display their result such as recommendations, prediction, real-time data, machine's condition, etc., through AR and VR solutions. The building blocks in blue, in Fig. 1, represent complementary services designed and developed as part of the SERENA project.

SERENA system and related to facilitating maintenance operations include eight data analytics related services, two running at an edge node, including the smart data generation at the gateway and the low-cost measurement service, and six intended for being executed at a cloud node, indicated with blue colour. Among the six analytics services in the cloud, four of them exploit data-driven approaches and are represented with darker blue in Fig. 1. One is mainly based on physical modelling. The last one performs the scheduling activity assigning maintenance activities dictated by the RUL estimation to the suitable shop-floor resources.

Besides, there are three additional services, encapsulated in the operator support box in Fig. 1 to support maintenance personnel, in terms of instruction provision and

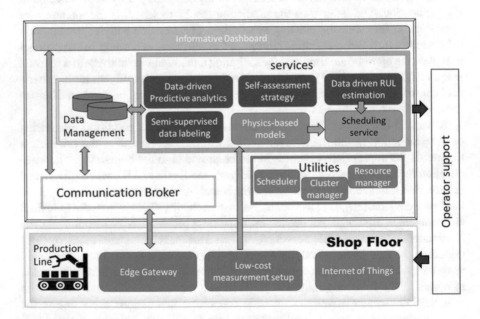

Fig. 1 SERENA user services

visualisation, along with an experimental service intended for personalised support, best fitting a worker's characteristics and skills. Finally, a central dashboard provides the SERENA system user with an overview of the supported per case functionalities and access to them.

The services mentioned above are described in greater detail in the following subsections.

3.1 Physics-Based and Hybrid Modelling (O&M Analytics Toolbox)

The physics-based modelling technique exploits a hybrid data analytics approach to perform conditional-based maintenance by combining data-driven and physics-based methods. In this context, the O&M Analytics Toolbox is a software library for implementing applications supporting industrial Operation and Maintenance. The needs and requirements for diagnostics and prognostics in the industry are many and varied. A toolbox, unlike an off-the-shelf application, facilitates addressing a wide range of those needs. On the other hand, even though the toolbox provides convenient building blocks for the work, building an application for a specific use requires programming. With the toolbox, this programming is done in the Python programming language. The toolbox is mainly intended for combining data analytics with domain-specific knowledge of, e.g., failure modes. Combining data-driven and physics-based modelling approaches is called hybrid modelling. Finding a cost-efficient balance between these modelling approaches depends on the application's business requirements and available data and knowledge.

The O&M Analytics Toolbox is built on top of Python Scientific Stack, a collection of open-source code libraries for mathematics, science, and engineering [15]. MIMOSA [16, 17] defines data structures and database schema specific for O&M, including, e.g., measurement metadata, information of the machines to be analysed, and diagnostic and prognostic results. These main software components are depicted in Fig. 2.

The implemented system indicates the condition of a monitored asset, with warnings of approaching faults. The system is implemented with low-cost hardware, sensor, and single-board edge computers to facilitate cost-efficient scaling to many assets.

3.2 Data-Driven Analytics

The data-driven services integrated into the SERENA Platform are the following:

- *Semi-supervised data labelling*
- *Predictive analytics services*

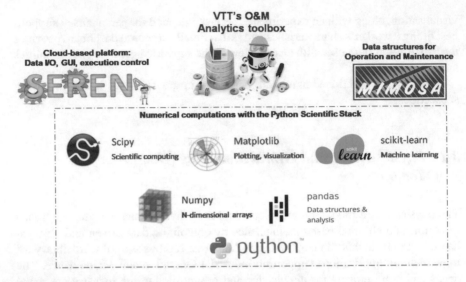

Fig. 2 O&M analytics toolbox

- *RUL (Remaining Useful Life) estimation*
- *Self-assessment strategy*

All the above services are enriched with a custom self-tuning strategy to offload the data scientist to manually set the specific algorithm parameters and efficiently exploit the benefits without expertise in the algorithms themselves.

Among the developed services, the predictive analytics service includes two main phases: (1) *Model building* requires a set of historical labelled training data to build a predictive model. Because this stage requires heavy computation resource consumption, it is usually performed on the cloud because it is, typically, not a heavy consumer of computational resources. (2) *Real Time predictions*. Once created, the model is applied in real-time to the new signals received from the industrial processes. This step can be performed either in the cloud or on edge. All the other services run on the cloud since they require managing a massive amount of historical data.

Services provide complementary features and address specific requirements characterising the Industry 4.0 applications. The company, based on its needs, might select only some of them. All services require a set of tasks to be carried out before exploiting data-driven services. Figure 3 shows the main analytics related functionalities integrated into the SERENA platform and detailed in Chap. 3.

Pre-processing and transformation tasks are mainly devoted to preparing raw data to be fruitfully exploited by data-driven services. The first includes (i) *Outlier and noise detection* to identify unwanted background values that affect the signals and generally lower the quality of the data-driven models. (2) Cycle alignment is an essential task responsible for adjusting data observations. All the collection points

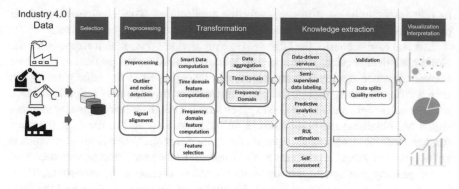

Fig. 3 SERENA data analytics tasks

have the same shape that may be required by the algorithms exploited in the analytics process.

The data transformation step includes various tasks mainly devoted to computing the most relevant features that accurately model the data under analysis. It comprises four main phases: (1) *Features computation* to extract critical segments based on a hybrid approach data-driven and domain knowledge, (2) *Time domain feature extraction* based on the computation of time-based features on signal segments to capture the variability of the signal better. Based on the predictions, a further transformation step may be required to model the signal evolution over time. (3) *Frequency domain feature extraction* computes features in the frequency domain. For example, time series is described as a sum of sinusoidal components (harmonics), e.g., using the Fourier Transform, and the most significant frequencies are kept into account for the analysis. (4)*Features selection* chooses only the most uncorrelated features to be used for the subsequent analytics steps.

A validation step assesses the quality of the designed services by exploiting state-of-the-art quality metrics based on the analytics task performed by the services.

More details on the methodology of the data driven analytics are provided in Chap. 3, while here the two main applications are detailed, namely prediction at the edge and RUL estimation, running at the edge and in the cloud respectively.

3.2.1 Real-Time Prediction at the Edge

The predictive analytics at the edge represents a core component since they fulfill different Industry 4.0 requirements and/or constraints. In some cases, the production process requires to have near real-time results from the analytic services. Thus, for all the applications that rely only on the predictive model's information and that do not require access to historical data stored in the cloud to perform a prediction, the trained model and its predictor can be deployed at the local gateway. In this way, the analytic component can analyze the input signals directly on the edge so that its outcomes can be used in near real-time applications. The latency of the whole process is drastically reduced with respect to the one deployed on the cloud.

In a real case scenario, the edge analytic prediction service, as it happens for the cloud analytical service, is wrapped in a docker container. In particular, the same docker container, deployed in the cloud, can be deployed in the edge gateway but with a different configuration: the prediction service on edge runs a Spark instance with a local design, and the service exposes REST APIs that allow ingesting an input signal and obtaining the predictions as an outcome. In practice, the edge gateway, through Node-Red, maintains a queue of incoming signals. When Node-Red collects an incoming message, it performs a POST request to the prediction service deployed in the gateway and synchronously waits for the prediction to be computed. Thus, the edge prediction service adds both the intelligent data computation and the forecast for the input signal, exploiting the predictive model and returning the Node-Red instance outcomes. Moreover, the prediction service on edge performs the intelligent data computation and the prediction for multiple incoming signals simultaneously paralleling the tasks. Once Node-Red has collected the prediction service results, it wraps the signals and the predicted values in the JSON-LD format, sending everything to the cloud. It is also possible to manage different versions of the predictor service on edge, exploiting the docker's versioning features. The trained model and the predictor service are wrapped together in the same container; in case of model updates, a new container will be created, and, if requested, it can be deployed from the cloud to the edge.

3.2.2   A Data-Driven Estimation of RUL

The RUL of an asset can be defined as the time left to the end of its expected operational life. Different degradation levels can characterise the same asset operating under different conditions, resulting in varying RUL values at the same point in time. Consequently, creating a generic model, statistical or otherwise, for estimating the RUL value is exceptionally challenging. Nevertheless, the RUL value's accurate estimation is a crucial component towards enabling a system's predictive maintenance. Thus, several efforts are underway to provide a solution to this problem. However, state-of-the-art algorithms, such as linear regression models, Hidden Markov Models, or Long Short-Term Memory Neural Networks, require extensive and complex modelling and training to estimate the RUL value with limited efficiency.

To that end and towards overcoming the limitations above, we propose data exploitation, collected out of different phases of operational asset life, to model its behaviour at specified and labelled timestamps and support the identification of deviations from its nominal or ideal functional profile and other tagged profiles. More specifically, data acquired at the deployment of an asset on a shop floor can be considered a perfect operational profile from which degradation gradually deteriorates its functional behaviour. Afterwards and during its usage for production purposes, data are used to estimate the deviation from this nominal profile. The distribution of the minor set of critical variables is calculated. Each selected variable's probability of belonging to the distribution is estimated using the Gaussian kernel density estimation. This probability representing the two profiles' deviation can then be associated with the RUL indicator or similar, facilitating its assessment. The lower the likelihood of each selected variable belonging to the data distribution of the machinery's correct functioning, the faster the degradation of object functionality. Thus the lower the RUL.

Based on the service mentioned above, the SERENA system can estimate the RUL value of an asset, machine, or robot. Moreover, considering a human-centered cyber-physical production system, the proposed approach supports its customized tuning and configuration by the human user towards addressing different scenarios and assets. It should be noted that an essential aspect of the presented method and as part of the implemented SERENA system is the knowledge of the domain expert for configuring the system and identifying the essential features for the analysis. This knowledge is considered part of the experience acquired during production and not knowing what an equipment manufacturer could provide unless monitoring the asset during its operational lifetime.

The RUL prediction service runs jointly with the Prediction Service. For this reason, the same request performed to the Prediction Service triggers the RUL service as well. When running the RUL service in the cloud, it monitors a target folder on HDFS, waiting for new incoming data. The service is triggered as soon as new data is collected and written by NiFi in the target HDFS folder. Then, every window of time (configurable as a parameter) performs the predictions for all the new signals, and it sends them back to NiFi, allowing other services to consume them and store them on HDFS.

3.3 Maintenance Aware Scheduling

In the SERENA project, a scheduling application has been implemented to support and improve the task distribution to human resources. SERENA Scheduling Tool consists of two main modules, the Scheduler Operating Module and the Scheduler Portlet Module.

3.3.1 Architecture

The scheduling tool's architecture is shown in Fig. 4 and includes two components. The first one, named the Scheduler Operating Module, provides the appropriate services to offer the optimised scheduling solution. Moreover, it communicates with the other parts of the SERENA platform and data management.

The second component, named the Scheduler Portlet Module, consists of multiple sub-components, which accomplish the interaction with the user and the functionalities provided by the User Interface (UI).

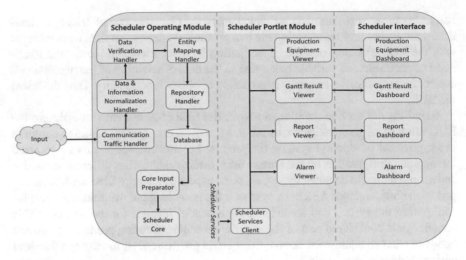

Fig. 4 Scheduling architecture

In particular, the Scheduler incorporates decision-making functionalities underneath to identify the best production-wise solution for scheduling the maintenance activities in line with the production plan. It focuses on specific constraints and generates multiple assignments of the tasks to the available resources. The challenge is to combine various criteria and try to fit the customer's needs to reach an optimised solution.

The weight of these criteria is getting defined and summed to calculate a utility value that estimates the customer's production targets, such as minimising the cycle time, production cost, energy consumption. Each weight ranges between 0 and 1. The process of identifying the best option of these normalised values is being performed inside the Scheduler Core. It contains an intelligent search algorithm to determine the best of the produced alternatives. This algorithm uses a three-parameter evaluation process to build its outcomes. These parameters are the Maximum Number of Alternatives (MNA), which defines the size of the grid; the Decision Horizon (DH), which represents the number of assignments that the algorithm will create; and finally, the Sampling Rate (SR) that defines the samples to be created. Subsequently, the algorithm's steps are described below:

1. Generation of the alternatives. One task is assigned to each available resource at a time, which can be easily adapted in the future.

2. Choose the desired criteria. It is essential to pick the correct criterion because this will quantify if alternative provides a better solution than the other.

3. Calculation of the sequence of the alternatives based on the desired criterion.

4. Selection of the best alternative. The alternative with the highest utility value is chosen as the best solution.

The overall result of the SERENA Scheduling Tool is the selection of the best alternative. The purpose of combining predictive maintenance and scheduling is to ensure that the proposed actions from analytics do not interrupt production or interrupt as little as possible because the failures are predicted in advance, and scheduling can be implemented according to the RUL. This means that the solution includes the tasks assigned to specific resources alongside this assignment's timing and duration. This schedule provides maintenance tasks fitted to an existing production plant.

3.3.2   Development Details

The Scheduling Tool is a web application designed as a client-server system using Java, supporting the connection to a central database containing the data required for planning/scheduling. The server-side includes a multi-criteria decision-making framework. The main components used for the scheduling application are the followings:

- **A Glassfish web server**: It hosts the main part of the application and the Jenna TDB that contains a knowledge representation of the needed information. The hosted application provides the main user interface and is also responsible for the REST calls of external systems that provide information, such as the RUL.
- **A Tomcat web server**: It hosts the scheduling and Gantt scheduling part of the application. It connects to the application running in the Glassfish & Jenna Triple Database to access the knowledge representation of the needed scheduling information.
- **A Cassandra database**: It is used to store time-based information.

Finally, the communication of the Scheduling Tool with other services is accomplished via REST API. The prototype is deployed as a package of Docker containers and managed through the Docker swarm manager.

3.3.3   Interfaces

The SERENA Scheduling Tool, shown in Fig. 5, is designed to provide users a friendly and flexible interaction environment. The UI has been designed to display the subsystems' results and give the maintenance engineers access to be informed regarding the maintenance tasks, plan, and assign and monitor the maintenance activities. The Scheduling Tool is aware of the company where it is running, so it will only display and edit the interested-company data.

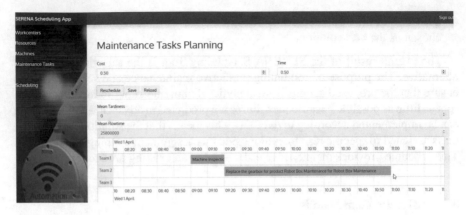

Fig. 5 Scheduling tool scheduling screen

3.4 Operator Support Services

Many companies continue to rely on traditional paper-based documentation to assist operators with maintenance tasks. They, therefore, fail to leverage the full potential of devices like mobile phones, tablets, or smart glasses to support their workers.

This batch of services focuses on providing support to the shopfloor personnel related to their maintenance activities. Machine measurements and maintenance instructions, including multimedia content, are supported and seamlessly provided via the implemented applications. As soon as a maintenance activity has been assigned to an operator, the operator support services are triggered. At the time of composing this document, the remote assistance and operator support tools include the followings:

- A notification hub for informing the operator about his/her assignment.

- A manual web editor, for authoring the instructions.

- A VR operator visualization app for real-time 3D animation of the machine superimposing valuable instructions, such as the analytics results.

- A personalisation module, via which an operator's feedback can be collected, evaluated and used to provide best-fit content over time.

The components mentioned above create a pipeline (Fig. 6), which facilitates dynamic operator support by providing multimedia content. A task assignment triggers the whole flow. The operator receiving the instructions on how to execute the assigned task can give feedback to the system. Then the user's input can either be stored or analysed by the personalisation service.

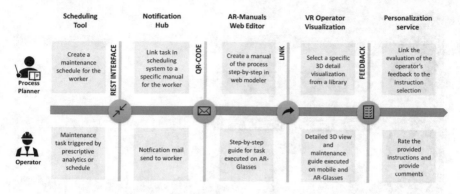

Fig. 6 Operator support pipeline

3.4.1 Main Dashboard

The main dashboard of the SERENA system, Fig. 7, is a unified and general-purpose UI. A SERENA user may access supported services while viewing the assets being monitored and related information. The dashboard is a web browser component integrated into the general SERENA Human-Machine Interface. According to the MIMOSA metadata model, all the 3D visualization services have been organized and integrated effectively and intuitively.

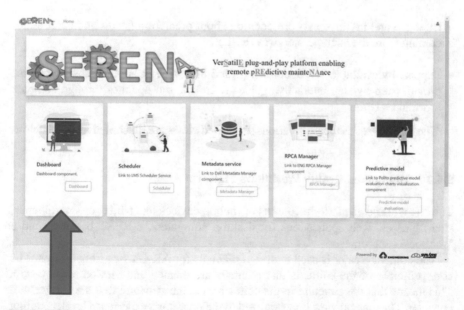

Fig. 7 SERENA dashboard components

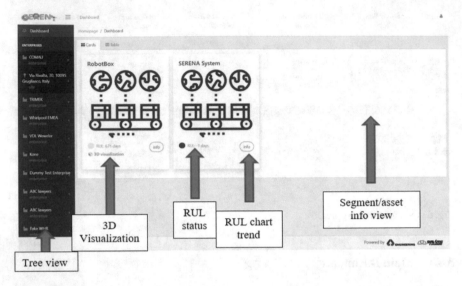

Fig. 8 Dashboard views

In particular, the dashboard facilitates the followings:

- Visualisation of the company facilities and its production systems as a tree view control with nodes based on the MIMOSA model, i.e., enterprise, sites, segments, and assets.

- Effective and intuitive data presentation about prediction (status and chart trend) coming from the analytics services (RUL).

- Support by available 3D models integrated into VR/AR applications which also provide step-by-step interactive guides to inform the operator through complex procedures better

  The SERENA dashboard's various supported views are illustrated in Fig. 8 below.

3.4.2 Augmented Reality Workflows Platform

The AR workflows platform consists of mobile devices like tablets or smartphones, smart glasses, web applications for desktop computers, a REST backend, and a WebSocket server.

   The content creator is used to design manuals for AR-Glasses from scratch. The core principle of the editor is that manuals are created and viewed step-by-step. This means that the structure in the editor is also the structure that the worker will consider. The manual view is customized to the operator's experience level, enabling more detailed explanations for untrained workers. The user flows to create a fresh

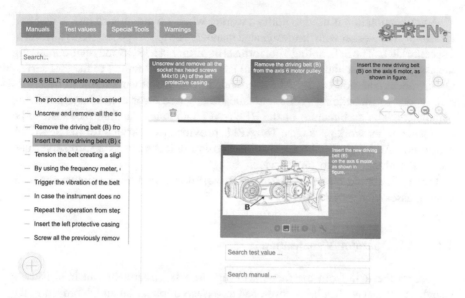

Fig. 9 Web modeler with image and text templates

step-by-step manual starts with editing the structure of the linear steps. Each step can then be revised in detail. The details may include textual instructions but also media playback for images and videos. Instructions may reuse similar elements, e.g., safety instructions or tool overviews from a general library, Fig. 9.

The worker may also record media (images, video) to document process results. In the future, with improved image recognition and processing models, the media feedback may also be machine processed to generate insights such as the evaluation of the correct manual execution.

The Manual Web Editor is a web application. The application frontend has been implemented in PHP and CSS. Actions in the application are handled through https-requests. The application is currently used as a prototype, with further development required to create a commercially usable application.

The workflows module can be deployed in different configurations based on the requirements of the use case partner. The default option for partners, which is preferred due to easy setup, is cloud deployment. The current platform used by the research partners for the demonstrator for testing, and improvement is deployed on Microsoft Azures Servers. After the testing and improvement phases, a platform may be deployed for each partner if required.

Due to the flexible deployment of the AR workflows platform together or without the SERENA swarm, it would also be possible to deploy the SERENA swarm on-premise and the AR workflows module in a cloud instance. This may be required for use cases where devices of external service technicians are not permitted in an on-premise network and therefore need a cloud-hosted example of the AR workflows module.

A notification hub is used to notify a worker when a maintenance task is required and supply the worker with the designated manual. The tool exposes a REST server at a specific URL. The Scheduling Tool formats its output in a JSON message containing the assignments' list and sends the result using the POST method to the tool server. Online documentation on the API is provided. This documentation also describes error codes for debugging API communication. These error codes handle missing parameters. The architecture of the REST-API facilitates the Interface also to be used by all other work packages. The API is provided as a service for the SERENA ecosystem. Additionally, the notification hub is implemented as a web server and hosted at a global web hosting service.

Finally, it should be noted that the AR workflows module is suitable for web, smart glasses, and mobile devices.

3.4.3   VR Operator Visualisation

To present the data from the manufacturing process effectively and intuitively, a visualization service has been developed as a SaaS application and deployed in the SERENA cloud. The visualization service can be considered a set of information pages that integrate machine data and plant information with the predictive analysis results. The visualization service consists of different tools depending on the service scope:

- A real-time 3D visualisation of the movement of the machinery with its status information.

- A real-time data are plotting visualisation.

- A visualisation of the information produced by the analytic services.

- A 3D virtual maintenance guide.

The aim is to support the maintenance and production engineer, giving them an effective tool to remotely evaluate the manufacturing process's status and guide the maintenance operator during the maintenance activities within the factory.

The service is implemented as an internal web app; it retrieves information from multiple sources in the SERENA platform using various protocols, such as REST and web sockets, aggregates and presents them to the user through a web browser.

The 3D visualization has the purpose of showing and present the real-time and prediction data to the operator in an intuitive and effective way.

The workflow to create this kind of visualization begins with a simplified 3D CAD model export of the involved machinery, porting it from the "vector" system to the "rendering" one, to make the application as light and usable as possible. In this way, the CAD model's engineering information is lost, so the VR/AR scene can be distributed without reverse engineering risk.

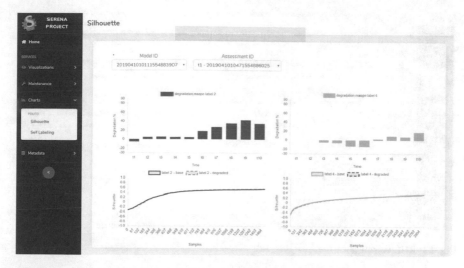

Fig. 10 Analytics visualisation

Fig. 11 Virtual maintenance interactive procedure

This service's main objective is to present the information coming from the manufacturing process directly on the 3D model in an interactive way, integrating them with virtual procedures that can guide the operator during the maintenance activities. Being a web service, it can be used on the move within the factory, with mobile and wearable devices. Some displays of the visualisation service are provided hereafter (Figs. 10, 11, 12 and 13).

Fig. 12 RUL visualisation

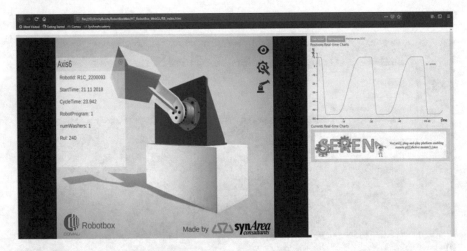

Fig. 13 RobotBox real-time visualisation

3.4.4 Personalised Support Service

This development's purpose is the personalised support of an operator in run-time by selecting the "best-fit" instruction set for an assigned task. Thus, the component creates a profile per user that is dynamically updated during task execution by his/her feedback and system collected measurements such as cycle time, waiting time, inter-action with the app, etc. The "best-fit" instruction is selected through a two-level collaborative filtering recommendation system. A high-level representation of the proposed workflow is provided below in Fig. 14.

The K-Nearest Neighbors (KNN) algorithm has been adopted to retrieve similar user profiles from a repository evaluating the distance between their characteristics,

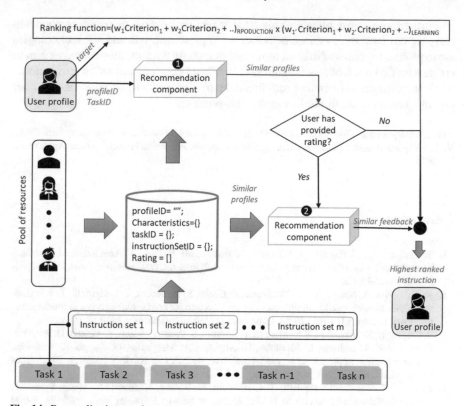

Fig. 14 Personalisation service concept

such as age, body type, experience, skill-set, etc. The KNN was selected as it is non-parametric, and training can be performed on the test dataset directly. The similarity of the available profiles' is evaluated by calculating the Pearson correlation. The output of the algorithm is a list of the operators' profiles with similar characteristics to the target user.

In particular, KNN modules were developed using the TensorFlow library and python 3.6. For data storage purposes a Cassandra database has been adopted, to facilitate future investigations on large datasets along with experimentation on different machine learning techniques. Data exchange between the back-end and the front-end is facilitate using RESTful services.

4 Conclusion

In this chapter we described the primary services exposed through the SERENA platform, covering the two main predictive analytics functionalities, detailed in Chap. 3, along with a scheduling tool consuming the predictive analytics outcomes.

Furthermore, two applications intended to facilitate the maintenance personnel in carrying out their maintenance activities were presented. AR and VR technologies are exploited to address different industrial needs. As a result, those services aim to create a unified batch of functionalities for a human-centred production environment while focusing on maintenance activities. The platform with its services can be either deployed on a private or public cloud or on-premise.

Acknowledgements This research has been partially funded by the European project "SERENA_ VerSatilE plug-and-play platform enabling REmote predictive mainteNAnce" (Grant Agreement: 767561).

References

1. T. Cerquitelli, D.J. Pagliari, A. Calimera, L. Bottaccioli, E. Patti, A. Acquaviva, M. Poncino, Manufacturing as a data-driven practice: methodologies, technologies, and tools, in *Proceedings of the IEEE* (2021)
2. N. Nikolakis, A. Marguglio, G. Veneziano, P. Greco, S. Panicucci, T. Cerquitelli, E. Macii, S. Andolina, K. Alexopoulos, A microservice architecture for predictive analytics in manufacturing. Proc. Manuf. **51**, 1091 (2020). https://doi.org/10.1016/j.promfg.2020.10.153
3. S. Panicucci, N. Nikolakis, T. Cerquitelli, F. Ventura, S. Proto, E. Macii, S. Makris, D. Bowden, P. Becker, N. O'mahony, L. Morabito, C. Napione, A. Marguglio, G. Coppo, S. Andolina, A cloud-to-edge approach to support predictive analytics in robotics industry. Electronics (Switzerland) **9**(3) (2020). https://doi.org/10.3390/electronics9030492
4. S. Proto, F. Ventura, D. Apiletti, T. Cerquitelli, E. Baralis, E. Macii, A. Macii, Premises, a scalable data-driven service to predict alarms in slowly-degrading multi-cycle industrial processes, in *2019 IEEE International Congress on Big Data, BigData Congress 2019, Milan, Italy, July 8-13, 2019*, ed. by E. Bertino, C.K. Chang, P. Chen, E. Damiani, M. Goul, K. Oyama (IEEE, 2019), pp. 139–143
5. F. Ventura, S. Proto, D. Apiletti, T. Cerquitelli, S. Panicucci, E. Baralis, E. Macii, A. Macii, A new unsupervised predictive-model self-assessment approach that SCALEs, in *Proceedings— 2019 IEEE International Congress on Big Data, BigData Congress 2019—Part of the 2019 IEEE World Congress on Services* pp. 144–148 (2019). https://doi.org/10.1109/BigDataCongress.2019.00033
6. D. Apiletti, C. Barberis, T. Cerquitelli, A. Macii, E. Macii, M. Poncino, F. Ventura, *ISTEP, An Integrated Self-tuning Engine for Predictive Maintenance in Industry 4.0* (IEEE, 2018), pp. 924–931
7. N. Nikolakis, A. Papavasileiou, K. Dimoulas, K. Bourmpouchakis, S. Makris, On a versatile scheduling concept of maintenance activities for increased availability of production resources. Proc. CIRP **78**, 172 (2018). https://doi.org/10.1016/j.procir.2018.09.065
8. N. Nikolakis, K. Sipsas, P. Tsarouchi, S. Makris, On a shared human-robot task scheduling and online re-scheduling. Proc. CIRP **78**, 237 (2018). https://doi.org/10.1016/j.procir.2018.09.055
9. G. May, M. Taisch, A. Bettoni, O. Maghazei, A. Matarazzo, B. Stahl, A new human-centric factory model. Proc. CIRP **26**, 103 (2015). https://doi.org/10.1016/j.procir.2014.07.112
10. D. Mourtzis, V. Zogopoulos, I. Katagis, P. Lagios, Augmented reality based visualization of CAM instructions towards industry 4.0 paradigm: a CNC bending machine case study. Proc. CIRP **70**, 368 (2018). https://doi.org/10.1016/j.procir.2018.02.045
11. D. Mourtzis, F. Xanthi, V. Zogopoulos, An adaptive framework for augmented reality instructions considering workforce skill. Proc. CIRP **81**, 363 (2019). https://doi.org/10.1016/j.procir.2019.03.063

12. A. Stratos, R. Loukas, M. Dimitris, G. Konstantinos, M. Dimitris, C. George, A virtual reality application to attract young talents to manufacturing. Proc. CIRP **57**, 134 (2016). https://doi.org/10.1016/j.procir.2016.11.024

13. N. Nikolakis, I. Stathakis, S. Makris, On an evolutionary information system for personalized support to plant operators. Proc. CIRP **81**, 547 (2019). https://doi.org/10.1016/j.procir.2019.03.153

14. N. Nikolakis, G. Siaterlis, K. Alexopoulos, A machine learning approach for improved shop-floor operator support using a two-level collaborative filtering and gamification features. Proc. CIRP **93**, 455 (2020). https://doi.org/10.1016/j.procir.2020.05.160

15. SciPy.org. Scientific computing tools for python. https://scipy.org/about.html

16. MIMOSA. www.mimosa.org

17. R. Salokangas, E. Jantunen, M. Larrañaga, P. Kaarmila, Mimosa strong medicine for maintenance, in *Advances in Asset Management and Condition Monitoring* (Springer, 2020), pp. 35–47

Industrial Use-cases

Predictive Analytics in Robotic Industry

Simone Panicucci, Lucrezia Morabito, Chiara Napione, Francesco Ventura, Massimo Ippolito, and Nikolaos Nikolakis

Abstract Modern breakthroughs have enabled the massive use of automation for addressing the ever changing automation market demands. To address an ever changing market and volatile requirements, every production process should be, if possible, optimised, reducing costs and wastes. At the same time, reconfiguration enabling flexibility through software automation and, in turn, resiliency is much pursued in the context of advanced manufacturing. One sector where automation is widely and deeply applied is the automotive one. For many years, the number of cars globally sold per year is above the 50 million at production rate up to 60 jobs (i.e., cars) per hour and even more. Therefore, even a rare and unexpected production stop has a significant financial impact in terms of production losses for the manufacturer. That is why predictive maintenance is the key aim of the SERENA project. The purpose of this chapter is to present the SERENA approach in the context of a robotic test-bed related to the automotive sector. As described in this chapter, the approach proposed is able to predict the Remaining Useful Life of an industrial robot key component, propagate that to the maintenance schedule system and the visualisation tool (including augmented and virtual reality maintenance procedures as well). The

S. Panicucci (✉) · L. Morabito · C. Napione · M. Ippolito
COMAU S.p.A., Via Rivalta 30, 10095 Grugliasco, TO, Italy
e-mail: simone.panicucci@comau.com

L. Morabito
e-mail: lucrezia.morabito@comau.com

C. Napione
e-mail: chiara.napione@comau.com

M. Ippolito
e-mail: massimo.ippolito@comau.com

F. Ventura
Department of Control and Computer Engineering, Politecnico di Torino, Turin, Italy
e-mail: francesco.ventura@polito.it

N. Nikolakis
Laboratory for Manufacturing Systems & Automation, University of Patras, Patras, Greece
e-mail: nikolakis@lms.mech.upatras.gr

© Springer Nature Singapore Pte Ltd. 2021
T. Cerquitelli et al. (eds.), *Predictive Maintenance in Smart Factories*,
Information Fusion and Data Science,
https://doi.org/10.1007/978-981-16-2940-2_5

entire software system is designed to run on a containerised infrastructure, providing thus flexibility, scalability and fault tolerance.

1 Introduction

The combination of information and communication technologies, together with automation technologies, the Internet of Things and services, makes higher and higher levels of integration within production, as well as between supplier and end customer possible today. Physical assets (e.g. Robots, CNCs, AGVs, welding and bonding systems, and many others) and software systems are increasingly interconnected. An increasing amount of machinery is now equipped with intelligent sensors, actuators and connectivity capability. The real-time availability of all relevant information from those physical components allows analysing and improving key aspects of product development, maintenance [1–3], logistics and other services. This new digitally interconnected equipment, delivers real-time production data where and when needed, helping in reducing downtime while improving overall quality. The robotics-oriented use case that is presented in this chapter describes how real-time analytics can provide customers with insight regarding robots' health status. Such information is made available locally and remotely using tablets and mobile devices. The evolution of robots into intelligent connected devices is radically reshaping factory and production. While individual measurements readings are valuable, factory often can extract powerful insights by identifying patterns in thousands of readings from many equipment over time [4]. In this use case for example, information from different individual measurements, such as a robot's motors temperature, vibration, and energy consumption, can reveal how performance correlates with the robot's engineering specifications. Information coming from measurements, such as heat and vibration, will show how we can predict an imminent bearing or belt failure days or weeks in advance, but also process quality nonconformity. Often referred to as big data and manufacturing intelligence, the analysis of such information and its distribution throughout the entire organisation provides customers with a powerful predictive and preventative maintenance tool.

2 SERENA Contribution

2.1 Architecture

The proposed SERENA architecture has been designed and developed aiming for resilience, both in terms of fault tolerance and scalability (horizontal and vertical). The SERENA system has been developed following a micro-service approach and with a generic enough data model addressing versatile industrial use cases.

All those specifics have been satisfied thanks to some key technologies, such as Docker Swarm, distributed storage and file system and the use of the MIMOSA standard as a data model. From the COMAU business point of view, the afore-mentioned design choices make the platform flexible enough to be integrated with property services, developed by COMAU, as well as some other applications and software architectures used by customers. As further explained in Sect. 2.4, in the scope of the COMAU use case, the entire system has been installed on premises. Currently, the system installed contains all the SERENA services and it is per-fectly integrated over the plant network and connected with the two machines used as test beds in the context of the project. The entire functional and data flows have been demonstrated and tested over time. They are made of the following steps:

- A physical gateway, which gathers raw data, creates signal features and the correct JSON format. Those operations are performed in a flow which is running on a containerised version of Node-Red.
- In addition to the data gathered in the previous item, there is another service which sends in real time the current positions of the machine to the central MQTT broker.
- Another service that is running on the gateway is the prediction service, which is based on a model trained in the central platform.
- Once a new machine cycle is ended and a new JSON produced, the data is sent to the Ni-Fi central broker, which is in charge of storing data both in the metadata repository (the MIMOSA storage) and the Hadoop distributed file system.
- In the central system, the visualisation service shows the actual status of the machine, the Remaining Useful Life (RUL) predicted, the digital twin of the machine and the replacement of the belt of the test bed with a procedure developed in virtual reality.
- In addition, the analytics service is able to give an output regarding the self-assessment (which says if the model performance is downgrading, suggesting a new training), self-labelling (to help experts in creating new and previously unknown machine behaviours) and the historical value of the Remaining Useful Life over time.
- At the end of the functional pipeline, the scheduling service takes the Remaining Useful Life prediction and organises the maintenance activities correspondingly.

For the time being, the SERENA system has proven good availability and reliability. Although it has been stressed with some network losses, hard restart of machines and power failures as well, it has given the expected response and resilience. At the time of writing, many of the SERENA architectural ideas as well as analytics procedures have been leveraged for the future releases of the COMAU IIoT platform, demonstrating the level of satisfaction with the outcomes of the project.

2.2 Approach

The COMAU use case aims to demonstrate the importance of predictive maintenance in term of business. Real equipment has been used in order to simulate, as well as possible, a real scenario. Nowadays a lot of money is spent for preventive maintenance as reducing the number of ordinary interventions does not represent a solution for savings, since the result would only be an increment of failures and down-times, causing a decrease of the production. Predictive maintenance can be a solution to this problem because it allows to monitor machinery in real time [5]. In this way, the manual inspections can be reduced without causing more production stops because all the equipment is under control. The proposed test-bed is a practical application of what has just been said.

2.2.1 Equipment Description

Typically industrial robots have up to 6 (and even 7) axes, most of which have a transmission belt to connect the motor to the respective gearbox reducer. Currently an operator, at periodic intervals, has to manually test the belt tensioning of each robot axis since their status and behaviour are a key factor of the correct robot functionality, both in terms of absolute precision as well as repeatability. This test-bed aims to demonstrate that machine learning can provide models able to figure out from themselves if the belt has some tensioning problems and send an alarm if it detects any.

More in detail, the purpose is to remotely classify the belt tensioning through a signal provided by the robot: the current absorption of each joint. Since tampering with an entire robot is very expensive, a test bed with the sixth axis of a COMAU industrial robot has been made. This is a test environment but having used real mechanical components makes the experiment significant. Moreover, this solution allows to avoid environmental noise and to better study the correlation between the provided data and the belt tension level. The test bed hereinafter is referred as *RobotBox* and it consists of various elements (see Fig. 1):

- a motor;
- a gearbox reducer;
- a transmission belt between the motor and the gearbox reducer;
- an encoder mounted on the motor to read its real time position;
- a 5 kg weight used to stress more the engine;
- a slicer to change the distance between the motor and the gearbox reducer, which simulates the changing the belt tensioning;
- a millesimal dial gauge.

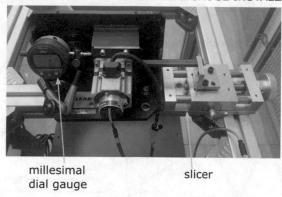

FRONT VIEW

BACK VIEW

5 kilos bulk

gearbox   belt   motor   encoder
reducer

BACK VIEW AFTER SLICER AND DIAL GAUGE INSTALLATION

millesimal   slicer
dial gauge

Fig. 1 Front and back view of *RobotBox* components used for the test bed

2.2.2 Data Acquisition

In order to set the experiment, it is necessary to define different belt tensioning levels. For this purpose, a slicer has been mounted on the test bed and it is able to linearly move the motor closer and further from the gearbox reducer. Then, a millesimal dial gauge has been installed to measure the relative displacement between motor and reducer. This is essential to be able to replicate each measurement and build a dataset with five classes with the same distance from each other.

The *RobotBox* completes a cycle each 24 s. In this time period it starts from the position of −500 degrees, then it slowly reaches the position of 90°, it stops for 5 s and it finally returns faster to −500°. Figure 2 shows the graph of the position during an entire cycle.

The signal of interest provided by the robot controller is the current absorbed by the motor to perform the cycle (the current signal is showed in the back graph of Fig. 3).

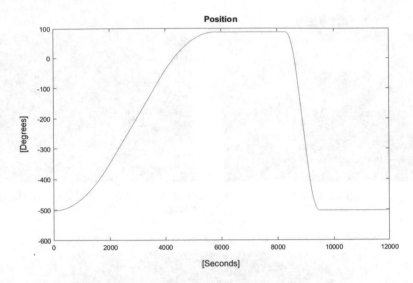

Fig. 2 Plot of *RobotBox* position signal belonging to an entire cycle

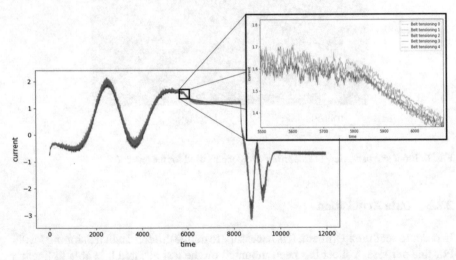

Fig. 3 Figure shows the absorbed current of one cycle of the *RobotBox*, collected with 5 different belt tensioning levels

In order to create a complete dataset, those signals have been collected for 5 different belt tensioning levels. Table 1 summarises what each class corresponds to.

In Fig. 3 a visual comparison of the current signals collected from each class is shown.

For each cycle, the number of data collected are around 12,000 as the time interval between two samples is 2 milliseconds (the sampling frequency 500 Hz).

Table 1 5 classes corresponding to the 5 different belt tensioning levels and number of samples for each class

Classes	Belt status	Number of samples
Class 0	Belt extremely tensioned	5952
Class 1	Belt too tensioned	1591
Class 2	Correct belt tensioning	1937
Class 3	Belt not enough tensioned	3174
Class 4	Belt with very low tensioning	4172

After having collected the raw data, all this information is bundled in a JSON file according to the MIMOSA schema and sent to the central services for the ingestion and processing.

2.2.3 Features Computation

In addition, COMAU as domain expert, provided some features in order to describe the phenomenon. These features are aggregate values calculated by the current signal of each cycle. They are:

- *Max*: maximum value.
- *Min*: minimum value.
- *Mean*: average value.
- *Median*: median value.
- *Std*: standard deviation value.
- *First quartile*: first quartile value.
- *Third quartile*: third quartile value.
- *Rms*: root mean square value:

$$x_{rms} = \sqrt{\frac{1}{n} \sum_{i=1}^{n} x_i^2} \tag{1}$$

where n is the number of points in the signal and x_i is the signal value of the i-th sample.

- *Kurtosis*: measure that defines how heavily the tails of a distribution differ from the tails of a normal distribution, calculated with:

$$x_{kurtosis} = \frac{1}{n} \frac{\sum_{i=1}^{n} (x_i - \bar{x})^4}{s^4} \tag{2}$$

where n is the number of points in the signal, $\bar{x}$ is the mean value and s is the standard deviation value.

- *Skewness*: measure of distribution symmetry, calculated with:

$$x_{skewness} = \frac{1}{n} \frac{\sum_{i=1}^{n}(x_i - \bar{x})^3}{s^3} \tag{3}$$

where n is the number of points in the signal, $\bar{x}$ is the mean value and s is the standard deviation value.
- *ElemOverMean*: number of points over the mean value.
- *AbsValueSum*: sum of absolute values of the considered samples, calculated as:

$$x_{AbsValueSum} = \sum_{i=1}^{n} |x_i| \tag{4}$$

- *AbsEnergy*: sum of the square values, calculated as:

$$x_{AbsEnergy} = \sum_{i=1}^{n} x_i^2 \tag{5}$$

- *MeanAbsChange*: the mean of the absolute value of $n - th$ discrete difference, calculated as:

$$x_{MeanAbsChange} = \frac{1}{n-1} \sum_{i=1}^{n-1} |x_{i+1} - x_i| \tag{6}$$

Features engineering Once the 14 features have been chosen, the current signal has been divided into 24 segments, as in Fig. 4, and for each of these time windows the aforementioned computation is performed. Then all of them have been calculated on each single segment, for a total of $14 \cdot 24 = 336$ features.

This strategy has been developed due to the fact that the differences between the classes could be more noticeable in specific parts of the signal.

Features selection Secondly, the number of features taken in consideration has been reduced using a technique based on Pearson correlation test [6] and a proper threshold. The remaining ones are the most significant. Removing some features affects the performance and the quality of the predictive model. In Sect. 2.3.1 the results of this step will be summarised in Fig. 5.

2.2.4 Data Analytics

In this robot industry use case, there was evaluated a subset of all the available analytic services developed for the SERENA platform, and mentioned in Chap. 3. In particular, predictive analytics, RUL estimation, self-assessment, self-labelling are the ones described below.

Predictive analytics The predictive analytics service, in this use case, evaluates three algorithms that have been available in the MLLib library [7], which are Decision

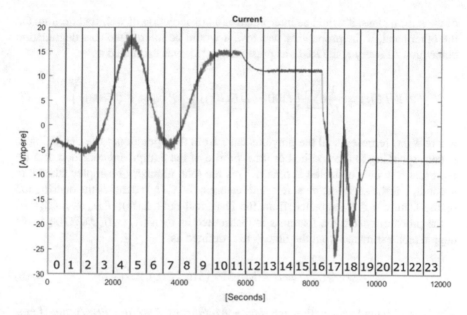

Fig. 4 Plot of the current divided by the 24 segments

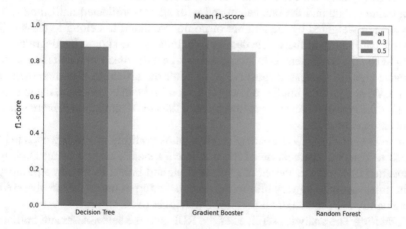

Fig. 5 Average f-measure for the different classifiers

Tree (DT), Gradient Boosted Tree (GBT) and Random Forest (RF). In particular, the service performs a grid search over the algorithms and a 3 fold cross-validation to test the configurations.

Remaining Useful Life (RUL) The RUL is the machinery degradation level estimation and it is calculated based on the most relevant available features. In order to evaluate the RUL, it has been supposed to have only the distribution of a baseline class of functioning. If a new observation had a high probability to belong to the distribution of the known class, it would be interpreted as an observation of well functioning. On the other hand, if the probability was low, it would be interpreted as a degradation behaviour. At every t, the RUL, as presented in [8], was estimated as

$$RUL(t) = \frac{1}{N_t} \sum_{x}^{X(t)} \left((100 - DEG(t)) \prod_{s}^{S} P(x_s \in K(X_s(t_0))) \right) \qquad (7)$$

where $X(t_0)$ represents all the previous data, from the operational life beginning to t_0 corresponding to the nominal or ideal profile of the equipment considered; $X(t)$ represent new data collected at time t; N_t is the total number of samples taken into account; S is the set of the most relevant features; $K(X_s(t_0))$ denotes the distribution of the feature $s \in S$ estimated from the historical data $X(t_0)$; $P(x_s \in K(X_s(t_0)))$ is the probability that the feature s is distributed as to $K(X_s(t_0))$; $DEG(t)$ is the degradation estimated from the time t_0 to t, defined as

$$DEG(c, t) = \alpha \times \text{MAAPE}(Sil_{t_0}, Sil_t) \times \frac{N_c}{N} \qquad (8)$$

which is obtained calculating the *Mean Arctangent Absolute Percentage Error* (MAAPE) [9] between the DS curves. In particular the one that take in consideration training data and the one calculated with all data collected until time t. This error is then weighted by the number of points predicted to belong to class c (N_c) w.r.t. the cardinality of the whole dataset (N). Finally, the sign to the degradation is identified by the coefficient α. In fact, this describe if the mean silhouette is increased (negative sign) or decreased (positive sign), after the arrival of new unknown data. So, a positive sign of α implies a lower cohesion in the data, since the mean silhouette after the arrival of new data is lowered and thus the degradation is increased (the degradation is positive).

Self-assessment The self-assessment service, whose validity has been proven in [10], affords to detect a degradation of the predictive model's quality using Descriptor Silhouette (DS) curve proposed in [11], as exploited index. In fact, in a production environment it could be very difficult, time consuming or unfeasible to observe and consequently collect data of all possible behaviours of machinery.

Self-labelling The analytics services of the RUL and the self-assessment both work assuming that the classes defined at the training phase are comprehensive of all the behaviours the industrial machine will exhibit. Nevertheless, that is a strict and not always feasible assumption, less than ever in the industrial scenario, where machine faults are rare and it is not possible to predict the entire fault set, due to the heterogeneity of the environmental conditions and processes. That was the reason why the service of self-labelling has been developed. Indeed, in case of the self-assessment warns that the model is loosing its effectiveness, the self-labelling service runs some unsupervised algorithms (and auto-tunes their parameters) in order to automatically suggest to a domain expert a first clustering of the data, showing as well the physical reasons, in terms of features, which have driven the underlying partition. Having this high-level perspective and system suggestion, it is easier for the domain expert to leverage his/her knowledge and label data, making the SERENA platform aware of the new classes. The model could be then re-trained with the novel information, if needed.

2.3 Results

From COMAU perspective, the results coming from the SERENA system can be seen from multiple points of view. On one hand, there is the business implication toward customers and, on the other hand, the outcomes of the different available services, tested with the *RobotBox*, for internal usage.

For the first aspect, the architecture designed in the project enables COMAU to be extremely flexible toward its customers. More precisely, the possibility to have both on premises and on remote cloud solution has made possible to reach an even wider set of customers and to became even more competitive in its reference market. In fact, the majority of the automotive players prefers a solution based on their own premises in order to protect their data from external attack but, on the other hand, small or medium enterprises, many of which do not have an internal ICT department, prefer to externalise the management of the cloud's resources.

Moreover, with micro-services oriented vision, COMAU has been able to easily build the applications and insert them in its portfolio and to maintain and deploy them to the customer in a very fast way. This architecture has made possible the scalability of the system based on customers' needs thanks to the Docker Swarm orchestration. This guarantees the reliability and resilience of the system and enables COMAU to be more competitive on the market.

Going into more detail about the services, the reminder of this subsection reports the results of the processing of data mentioned in 2.2.3 and the outcomes of the analytics methodology proposed in 2.2.4.

2.3.1 Features Computation Results

After the computation of the features and in order to consider only the most significant ones, the features selection step has been performed and it has turned out to be winner. In fact, it has positively affected the model's quality. As in Fig. 5, the f-measure (i.e., the harmonic mean of precision and recall [12]) obtained with a threshold equal to 0.5 for correlation, performed better for all the classifiers tested. Both the option with threshold equal to 0.3 and the option with all features selected performed worst.

2.3.2   Data Analytics Results

Predictive analytics

For this use case, the grid search methodology selects the best algorithm within Decision Tree, Gradient Boosted Tree and Random Forest. The choice was the Random Forest. Moreover, in Table 2 are reported the 10 most important features. These are selected calculating their importance values according to [13].

When the best model has been chosen and, based on the scenario, it is stored on HDFS or deployed on the edge. Consequently, real-time prediction both on the cloud and on the edge becomes available.

In this case, the deployment on the edge has been implemented. The gateway hosts both features computation and prediction of new data, maintaining a queue of current incoming signals. The tasks have been executed in parallel on a local deployment of the Spark service. The evaluation of the latency was done with a dual cores edge gateway with 4 GB of RAM and with 1000 signals in sequence evaluated one by one. The result of this test was an average computation time of 2.73 s with standard deviation equal to 0.36. This computation time is clearly acceptable, considering that a *RobotBox* cycle lasts 24 s.

Remaining useful life (RUL) Firstly, notice that the *RobotBox* current's signal has been divided into 24 segments. Thanks to domain experts knowledge, the most relevant ones have been selected in order to consider the ones in which robot activity is most challenging and consequently subjected to failures. These segments are the 10th and the 11th, just before the idle phase of the cycle. In order to evaluate the RUL, it has been supposed to have only the distribution for class 2 and 4 that represent a good functioning of the asset. Figure 6 shows the Gaussian Kernel density of the feature *mean*, for one of the previously mentioned two segments and for, on one hand, the known classes and, on the other hand, for the new class, the class 0, that represents the degradation of the system. To test this service, the injection pattern shown in Table 3 was used. In particular, at in every specific timestamp t_i has been

Table 2 Most important features selected from the best RF

Importance	Feature name	Segment number
0.048	skewness-1	1
0.048	mean_abs_change-6	6
0.045	mean_abs_change-1	1
0.044	mean_abs_change-5	5
0.040	mean_abs_change-4	4
0.038	mean_abs_change-10	10
0.030	kurtosis-3	3
0.029	mean_abs_change-7	7
0.023	kurtosis-1	1
0.023	third_quartile-10	10

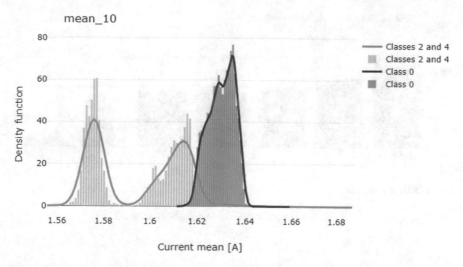

Fig. 6 Distribution of the mean of the current in segment 10

Table 3 Data injection pattern

t	Label 0	Label 1	Label 2	Label 3	Label 4
t_0 (training)	–	–	1500	–	3000
t_1	–	–	250	–	–
t_2	–	–	473	–	400
t_3	–	–	473	–	800
t_4	–	–	473	–	1172
t_5	-	-	473	-	1172
t_6	2000	–	473	–	1172
t_7	4000	–	472	–	1172
t_8	5952	48	473	–	1172
t_9	5952	1591	473	457	1172
t_{10}	5952	1591	473	3174	1172

added to the training data a number of samples for each label. Notice that, until time t_6 was not injected data of unknown classes. Figure 7 shows the results of the test and underlines that at t_6, when unknown data class was injected, the RUL percentage drastically decreased. Moreover, Fig. 8 shows the data degradation percentage that at t_6 increase significantly.

Self-assessment As mentioned before, it has been assumed that classes 2 and 4 were the known ones during the training of the algorithm. Then new incoming data of classes 0, 1 and 3 has been injected. To test the self assessment service a trigger was used based on incoming samples number in order to be time and process independent. The injection pattern is the same used for RUL estimation and is available in Table 3. In every specific timestamp t_i has been added a number of samples for each label. As

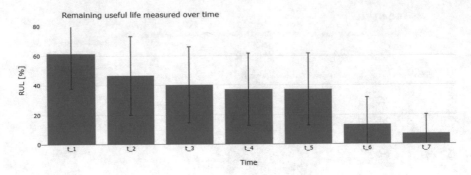

Fig. 7 RUL estimation

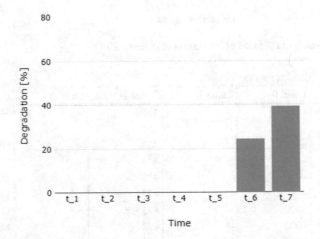

Fig. 8 Data degradation over time

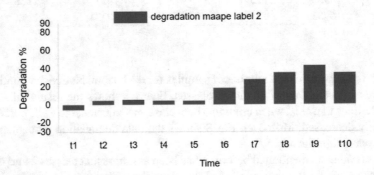

Fig. 9 Percentage of degradation for class 2 and 4

Best Algorithm

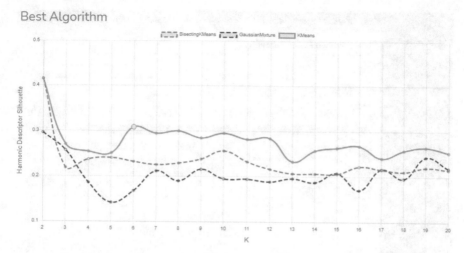

Fig. 10 Self-labelling training algorithm results

before, until time t_6 was not injected data of unknown classes and, in fact, Fig. 9 shows the degradation started to be significant at that timestamp. In fact, Fig. 9 shows the percentage of degradation of class 2 measured by the self assessment cloud service. From t_6 forward the degradation became more and more clear and was correctly identified a new and unknown distribution of data.

Self-labelling

Figure 10 shows the three unsupervised algorithms used by the self-labelling system (i.e., Bisecting K-Means [14], Gaussian Mixture [15], and K-Means [14]) and the evaluation performed on data in order to auto-tune the training parameters.

2.3.3 Scheduling and Visualisation Results

Scheduling

In order to close the loop of the predictive maintenance, the results of the analytics are used by a scheduling service able to plan the maintenance activities in order to prevent the failures. The user interface of the scheduling service is illustrated in Fig. 11.

In this use case, it was triggered by the RUL estimation and consequently every 11 ms a new schedule was generated. There are two different possible tasks (machine inspection and replacement of the gearbox) and three potential resources (one newcomer and one of middle experience, one newcomer and one expert or an expert). For each combination of task and resources are also available minutes needed and cost.

Visualisation Essentially this service makes available for the user, in a web interface, three different panels: visualisation, charts and maintenance. In the first is

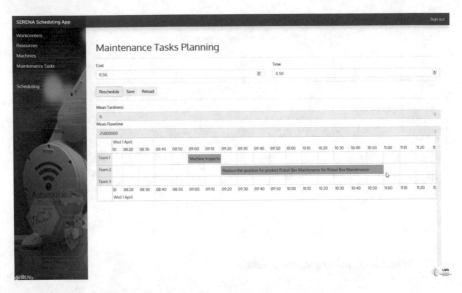

Fig. 11 Scheduler Gantt chart

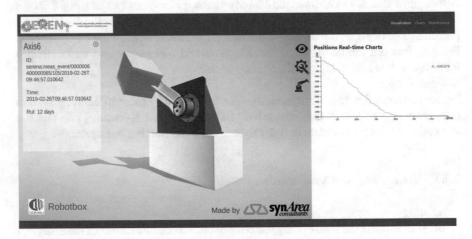

Fig. 12 Screenshot of visualisation service

visualised the 3D model of the *RobotBox* and its real movements. Moreover information about the prediction's results is available, for example a coloured label on the motor that shows its status. The chart panel shows in detail the results of the analytics service and finally the maintenance panel consists of a 3D guided procedure for maintenance operations. The user interface of the visualisation service is displayed hereafter in Fig. 12.

2.4 System Deployment

In this section, the on premises deployment of the reference SERENA architecture is going to be discussed. Automotive is, so far, the COMAU reference market since it is one of the main targets for automation due to its impressive volumes.

Even if there are some production plants where remote cloud solutions are completely accepted and also preferred, there are still numerous use cases or car manufacturers which prefer the on premises solution, defining that need as a requirement. That is why, in the scope of this project, in addition to the cloud and the hybrid-cloud proposals and demonstrations, the COMAU use case has been developed keeping as requirement a completely on premises scenario, in order to actually demonstrate all the different capabilities and versatility provided by the SERENA architecture.

Being flexible in the deployment it is worthwhile, not just from a technical point of view (since it requires to keep the architecture micro-service oriented as much as possible), but also from the business prospective. Indeed, with this kind of approach it is possible to satisfy a wide range of requirements coming from any kind of customers, starting from small ones to the biggest, like car makers.

As Fig. 13 shows, the SERENA system installed on COMAU premises includes all the platform services and has two *RobotBox* from which the SERENA gateway gathers data, before sending it to the central system. From the hardware perspective, the entire system has been installed on a machine with 16 cores and 32 GB of main memory. Currently, we have not started the installation of a distributed storage solution yet, but the activities are already planned. Nevertheless, the SERENA platform uses some Docker Swarm features to abstract from the specific storage solution

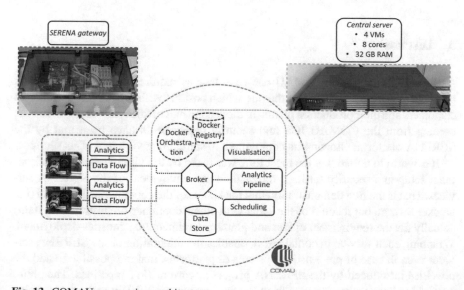

Fig. 13 COMAU on-premises architecture

adopted in order to easily change that and adapt to the one preferred or used by final customers.

The overall installation and integration activities took about one month and in general it seems reasonable to say it would take between one and two months to implement all the services and connect the system to the production machines or equipment. It is worth to highlight that the design and development of ad-hoc analytics algorithms is not taken into account in that deployment estimation. Summing up the on premises deployment, the SERENA architecture satisfied the COMAU security policies and it was fully integrated in the COMAU network, allowing to connect to machineries and equipment all over company plants. The security requirements were successfully accomplished by the fact that the entire SERENA platform is exposed to the COMAU network via a single-service point-of-access, which accepts only HTTPS connection requests coming from SERENA-certified hardware, managed by the own certification authority.

Some scripts have been developed to accelerate the deployment time and effort. Those are able to create the virtual machines requested by the distributed environments with the network and security setup, install all the software requirement for the micro-service architecture and the core SERENA services as the HDFS and MIMOSA storage, as well as the Ni-Fi broker and the RPCA.

This use case is just the on premises demonstration of the SERENA system, which offers a wide range of different deployment strategies in order to deal with a variety of customer requirements, keeping the same reference architecture. Concretely, that means that at services level, it does not matter which is the final environment or deployment, since all that complexity has been managed at architectural level, thanks to micro-services and container-orchestration system.

3   Discussion

As aforementioned, the COMAU use case has, as requirement, the complete on premises deployment, since it is the one which best fits with the biggest automotive customers and their production plants. In the second half of 2019, all the requirements coming from the COMAU ICT and security departments have been met by the SERENA platform, allowing its deployment to start, along with that of its services.

It is worth to notice that the most time was taken by the virtual machine environment setup and security settings, rather than the deployment of the SERENA services. That is the first demonstration that the choice of the containerised solution (i.e. Docker Swarm, but it could be the same with others container-orchestration system) actually fits the requirement of fast and platform/architecture agnostics deployment. Wrapping each service in container(s), enables the integration of the SERENA services even in case of pre-existing systems or platforms, making possible to add the novelties introduced by the SERENA project in current IIoT pipelines. The entire system has been up and running from the final months of 2019 up to the first months of 2021 and its resilient, fault tolerance and load balancing have been stressed and successfully tested.

Most of the know-how developed about the architecture and containerised systems has already been used internally and had a profitable impact on the current COMAU IIoT product and portfolio.

The architectural design and efficiency of the platform is, with any doubts, one of the best achievements of the SERENA project; nevertheless, the key enabler of the predictive maintenance is the analytics pipeline. From the beginning of the project, the objective regarding the COMAU use case was to train a supervised model able to predict the machine or industrial equipment RUL. In addition to the prediction of production line sudden stops (which is the main goal) the system has been designed to also propagate that failure prediction to all the stakeholders involved, such as maintenance crew, plant managers and control room operators.

As deeply described in Sect. 2.3, the analytics pipeline developed under the scope of the project has achieved the prediction objective, adding novel features which take the name of self-assessment and self-labelling. Those two additional services are in charge of, respectively, notifying if the current model is not able to successfully predict anymore, and suggesting a first way to cluster data coming from the actual field which are not recognised as ascribable to the classes which the model has been trained with. The model prediction degradation is a key forward-looking feature since industrial equipment (especially robots) may be used in a wide range of different environment conditions (such as temperature, humidity and so on), with different end-effectors and for dealing different objects (which both change the payload handled by the machine) and with and infinite number of different working cycles. That huge heterogeneity means that it is a really tricky challenge to create a dataset able to represent the entire distribution of the machine behaviour for any payload, any cycle and in any environment conditions. Having a system which firstly notifies you the need to re-train the model and then suggests as well a first cluster for the new data, allows to always have efficient models, in addition to the possibility to increase the prediction capabilities during the working life of the equipment.

The output of the prediction model, the RUL, is then used by the scheduling and visualisation service to effectively schedule the maintenance interventions and to show some maintenance procedures in augmented or virtual reality.

4 Outlook

The first thing that could seem expected is that two industrial machines, designed by the same company and produced with the same process and materials, may have the same behaviour and physical response if the environment conditions and working cycles are the same. Unfortunately, that can not be taken as an assumption. Indeed, although we have build two *RobotBox* with exactly the same components and in the same way, they do not have the same behaviour in terms of signals and features. This discouraging introduction is just to highlight the importance of having a more general and comprehensive dataset as possible in order to train predictive models which may bring effective benefits.

Thus, the main future development is to find a way to generalise the knowledge leverage by the SERENA project in order to effectively extend the predictive capabilities on an entire robot, since the belt tensioning is a key aspect in terms of precision and repeatability. The choice of starting on a single robot axis was mandatory since the signals extracted in case of entire robots contain information coming not only from the robot components and the environment condition, but also all the physical interaction due to the kinematic chain, which would be the worst case to start from.

Watching to the SERENA project as a comprehensive IIoT product, future developments should regard the visualisation service, since so far it was build to show the project results but it has never been tested in a real scenario and thus in terms of scalability (i.e. trying to add and manage thousands of devices), user experience, capability to show just the information needed depending on the user logged and users management.

So far all the features provided by the analytics pipeline have been presented and described but, apart from the prediction on novel data, all the procedures of training a new model, starting the self-assessment and the self-labelling are actually run manually. An interesting future development regards the design and development of a service which is in charge of triggering those functionalities, asking operators about information needed (or authorisations) and providing periodic reports and explanation about the actions taken. Doing so, it is possible to have a warranty that the prediction system is kept updated and what is the confidence level of the current model(s).

As aforementioned, the SERENA system is completely designed and developed in line with a containerised architecture, which offers, among others, the benefit of fast and flexible deployment on physical hardware, virtual machines and any kind of cloud. The last future development proposed is then to extend the existing deployment scripts, guided by a graphical user interface, to install the entire platform (and to choose services) in all the scenarios listed above, making the deployment even faster and self-managed by the platform.

Acknowledgements This research has been partially funded by the European project "SERENA – VerSatilE plug-and-play platform enabling REmote predictive mainteNAnce" (Grant Agreement: 767561).

References

1. D. Apiletti, C. Barberis, T. Cerquitelli, A. Macii, E. Macii, M. Poncino, F. Ventura, istep, an integrated self-tuning engine for predictive maintenance in industry 4.0, in *IEEE International Conference on Parallel & Distributed Processing with Applications, Ubiquitous Computing & Communications, ISPA/IUCC/BDCloud/SocialCom/SustainCom 2018, Melbourne, Australia, December 11–13, 2018*, ed. by J. Chen, L.T. Yang (IEEE, 2018), pp. 924–931

2. S. Proto, E.D. Corso, D. Apiletti, L. Cagliero, T. Cerquitelli, G. Malnati, D. Mazzucchi, Redtag: a predictive maintenance framework for parcel delivery services. IEEE Access **8**, 14953 (2020)
3. T. Cerquitelli, D.J. Pagliari, A. Calimera, L. Bottaccioli, E. Patti, A. Acquaviva, M. Poncino, Manufacturing as a data-driven practice: methodologies, technologies, and tools. Proc. IEEE **109**(4), 399 (2021). https://doi.org/10.1109/JPROC.2021.3056006
4. T. Cerquitelli, D. Bowden, A. Marguglio, L. Morabito, C. Napione, S. Panicucci, N. Nikolakis, S. Makris, G. Coppo, S. Andolina, A. Macii, E. Macii, N. O'Mahony, P. Becker, S. Jung, A fog computing approach for predictive maintenance, in *Advanced Information Systems Engineering Workshops—CAiSE, International Workshops, Rome, Italy, June 3–7, 2019, Proceedings, Lecture Notes in Business Information Processing*, ed. by H.A. Proper, J. Stirna (Springer, 2019). Lecture Notes in Business Information Processing, vol. 349, pp. 139–147
5. R. Pinto, T. Cerquitelli, Robot fault detection and remaining life estimation for predictive maintenance, in *The 10th International Conference on Ambient Systems, Networks and Technologies (ANT 2019) / The 2nd International Conference on Emerging Data and Industry 4.0 (EDI40 2019) / Affiliated Workshops, April 29–May 2, 2019, Leuven, Belgium, Procedia Computer Science*, ed. by E.M. Shakshuki, A. Yasar (Elsevier, 2019), vol. 151, pp. 709–716
6. S.M. Ross, *Introduction to Probability Models*, 6th edn. (Academic Press, San Diego, CA, USA, 1997)
7. X. Meng, J. Bradley, B. Yavuz, E. Sparks, S. Venkataraman, D. Liu, J. Freeman, D. Tsai, M. Amde, S. Owen et al., Mllib: Machine learning in apache spark. J. Mach. Learn. Res. **17**(1), 1235 (2016)
8. S. Panicucci, N. Nikolakis, T. Cerquitelli, F. Ventura, S. Proto, E. Macii, S. Makris, D. Bowden, P. Becker, N. O'Mahony, L. Morabito, C. Napione, A. Marguglio, G. Coppo, S. Andolina, A cloud-to-edge approach to support predictive analytics in robotics industry. Electronics **9**(3), 492 (2020)
9. S. Kim, H. Kim, A new metric of absolute percentage error for intermittent demand forecasts. Int. J. Forecasting **32**(3), 669 (2016)
10. T. Cerquitelli, S. Proto, F. Ventura, D. Apiletti, E. Baralis, Towards a real-time unsupervised estimation of predictive model degradation, in *Proceedings of Real-Time Business Intelligence and Analytics* (2019), pp. 1–6
11. F. Ventura, S. Proto, D. Apiletti, T. Cerquitelli, S. Panicucci, E. Baralis, E. Macii, A. Macii, A new unsupervised predictive-model self-assessment approach that scales, in *2019 IEEE International Congress on Big Data (BigDataCongress)* (IEEE, 2019), pp. 144–148
12. P.A. Flach, M. Kull, Precision-recall-gain curves: Pr analysis done right, in *NIPS*, vol. 15 (2015)
13. T. Hastie, R. Tibshirani, J. Friedman, *The Elements of Statistical Learning: Data Mining, Inference, and Prediction* (Springer Science & Business Media, 2009)
14. J. Hartigan, M. Wong, Algorithm as 136: A k-means clustering algorithm, in *Applied Statistics* (1979), pp. 100–108
15. P.N. Tan, M. Steinbach, V. Kumar, *Introduction to Data Mining* (Addison Wesley, 2005). URL http://www.amazon.com/exec/obidos/redirect?tag=citeulike07-20&path=ASIN/0321321367

Remote Condition Monitoring and Control for Maintenance Activities in Coordinate Measurement Machine: A Data-Driven Approach for the Metrology Engineering Industry

Andrea Gómez Acebo, Antonio Ventura-Traveset, and Silvia de la Maza Uriarte

Abstract Based on the high level of customisation and accuracy demanded in manufacturing processes, the increasing amount of data managed, and the expanding of metrology market along with the evident necessity of reducing maintenance costs while increasing efficiency in companies, SERENA remote predictive maintenance developments are implemented and validated in a metrology industry test bed that aims to monitor the performance of a Coordinate Measurement Machine (CMM) for maintenance process improvement. TRIMEK, supplier of a wide range of products and solutions for quality control and in-line inspection, has installed sensors in the most critical system of a CMM that impacts on the correctness of dimensional measurements: the air bearings, in order to gather information about machine state and exploit condition monitoring data to develop predictive maintenance tools for CMMs; additionally composed by a data analytic service for early anomaly detection, a scheduler system for maintenance activities and an operator support system to help employees to perform their maintenance tasks. This cloud-based general solution represents an improvement for the service and support provided by TRIMEK to its customers, that leads to the reduction of production breakdowns and maintenance costs, along with the improvement of efficiency, machine life-cycle and performance.

A. G. Acebo (✉) · S. de M. Uriarte
TRIMEK, Madrid, Spain
e-mail: agomez@trimek.com

S. de M. Uriarte
e-mail: smaza@trimek.com

A. Ventura-Traveset
TRIMEK, Barcelona, Spain
e-mail: toni.ventura@datapixel.com

1 TRIMEK Use Case

1.1 *Introduction*

All manufacturing processes require high accuracy. The actions of maintenance, calibration, performance evaluation and failure detection represent an important aspect of achieving this required accuracy, carrying costs for an enterprise. In this context, metrology equipment is essential for quality control and in-line inspection. Therefore, proper calibration and maintenance of metrology solution machines are considered crucial in the whole metrology process as well as in the manufacturing process, despite the fact that they are very stable machines.

TRIMEK (logo in Fig. 1) is one of the main manufacturers of metrological systems and solutions worldwide, and is the leading company in the Basque Country and Spanish markets in the field of CMMs along with measuring and digitalisation software, covering any metrological need that a company might have, e.g. dimensional measurements, metrology machinary calibration and maintenance. TRIMEK's systems are usually applied in industrial environments, automotive and aeronautics sectors.

1.2 *Coordinate Measuring Machine (CMM)*

Coordinate Measuring Machines (CMM), are dimensional inspection systems (devices) specifically designed for measuring the geometry of physical objects/parts and their digitisation in order to create their virtual image from the point clouds sensed. This is subsequently analysed with the M3 software (High Performance Software for capturing and analysing point clouds), obtaining the corresponding geometrical dimensions and deviations, based on the CAD model of the object/part. These systems use non-contact sensors or touch probes for objects' scanning through its displacement from a reference position in a three-dimensional Cartesian coordinate system (XYZ). In this sense, the three machine axes move independently of one another to enable the total coverage of the measuring volume.

The machine mainly consist of three different elements:

- The optical sensor which is in charge of obtaining high-accuracy 3D point-clouds representing digital images of the manufacturing parts. Part information is captured

Fig. 1 Trimek—Basque company for the design and manufacturing of coordinate measuring machines

using laser triangulation techniques and the selection of the specific optical sensor depends on the accuracy, speed, and fidelity required by the measurement. Touch probes may be considered as well.

- Mechanical parts: surface, arm, bridge, air bearings.
- Control system: in charge of moving the sensor within the measuring volume according to the trajectories, paths, positions, and orientations previously defined in the digitisation program and measuring software.

In the context of the SERENA project [1, 2], TRIMEK included a Bridge-type machine; in particular the SPARK model, that is manufactured with reduced complexity in order to keep it as robust as possible (see Fig. 2). The machine is suitable for highly accurate measurements, scanning and digitising tasks; achieves a high dimensional stability and precisely controlled movements. The measuring system of each axis is mounted on Robax (a thermally inert material) which combined with the granite construction, makes the SPARK an accurate and fast machine even under temperature variations.

The selected machine has the following main features:

- High precision XYZ structure.

Fig. 2 CMM brigde type machine—SPARK model

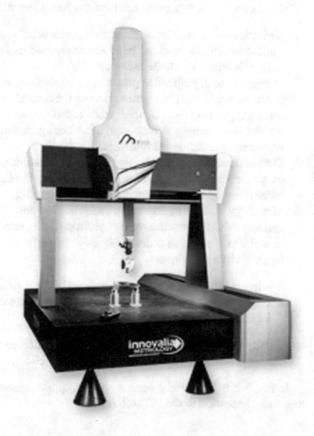

- High precision air bearings for best accuracy motion.
- Measuring range (in mm): X 500–3000; Y 500–3000; Z 600–2000.
- Standard accuracy MPE: 1,5 + 2,2 *L/1000 (TP2 or TP20, Manual or Automatic Temperature Compensation and Resolution of 0.0005 mms.)
- Max. Speed 3D: 500 mm/s.
- Max Acceleration 3D: 850 mm/sg.
- Air Consumption: 150 l/min.
- Air supply pressure: 6 bar.
- Electricity consumption: max 3000 VA
- Full PH10M, PH10MQ, PHS1 multi-wire integration.
- High-definition scanning.
- Combines touch technology for accessing difficult areas.

The SPARK machine is provided to TRIMEK's customers with the appropriate software tool to support its measuring operation, namely the M3 software, to capture and analyze point clouds and compare them with the CAD files in order to detect deviations. According to the measuring plan, its configuration is customized and defined (based on client's requirements) in order to collect key dimensional information generated in the measurement process.

The measurement process consists of the following steps:

1. Definition of clients' requirements and measuring plan.
2. Calibration of the machine once the requirements are established, in order to ensure the accuracy of the measures.
3. Design of the digitalization program by M3 Software, based on the definition of the movements (rotation, paths, etc.) of the optical sensor around the manufactured part that needs to be measured. Furthermore, the determination of sensor quality, the scanning settings, and other configuration settings of the measuring equipment are defined at this stage.
4. Data collection, structuring, and ordination: definition of the "reference point" of the machine. This reference is considered by the CMM as a"zero position" in order to set up the measurements in the 3D coordinates (XYZ) space.
5. Metrology analysis including the analysis of the part's point cloud captured by the machine sensor. Quantitative comparison of the point-cloud and the ideal CAD model is performed in order to identify dimensional deviations. This analysis is based on two techniques:

   - Alignment (to refer to the same axis origin both virtual parts and CAD models)
   - Color-mapping (evaluation of the distances between point pairs by applying a color code).

6. Finally, the M3 software stores all the information generated. Measuring reports can be provided/generated if required by the client.

In Fig. 3 it is indicated a high level conceptual representation of the measuring machine connection to the M3 software.

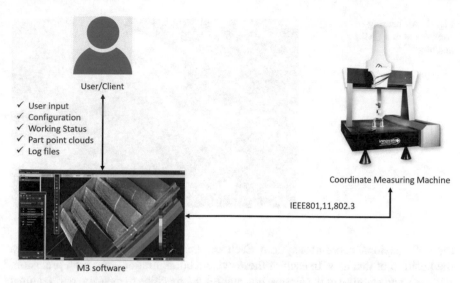

Fig. 3 High level conceptual representation of the measuring machine connection to the M3 software

The output from these machines requires a high degree of customisation as per the requirements of a wide variety of end-users and to guarantee that the physical product does not present any defect that might cause a problem or waste of money to our clients, it is key to ensure the required accuracy of the CMM in each measurement. This accuracy, along with the avoidance of unexpected breakdowns, depends on the performance of the machine and its components.

1.3 CMM Subsystem Under Study

In the group of components that form a CMM, one of the most critical is the air bearing system (see Fig. 4). This can be linear or rotary, and it floats on air above its guide ways, representing a highly stable non-friction system that plays a key role in delivering the precision required in metrology equipment. The most common causes of air bearings failure are air contamination and out of balance forces causing vibration.

Performance criteria, such as repeatability and minimized error motions, geometric precision straightness and flatness for linear bearings, are points in favour for the use of air bearings in applications where the ultimate precision counts, as is the case of metrology equipment. CMMs and metrology tools use a variety of design architectures, including moving gantries, static gantries, cantilevered X-Y-Z, and many others. The most common type of CMM, also represented in the machine selected for the testbed, is composed of three orthogonal axes (X, Y, and Z) operating in a

Fig. 4 Air bearing
component in a CMM
machine

three-dimensional coordinate system. Each axis has a position encoder that indicates
the position of that axis. In each of these system configurations, the high precision,
stiffness, and stability of the air bearing enables the machine to measure part features
with sub-micron accuracy and fidelity.

For this reason, based on TRIMEK personnel's experience the compressed air
system was selected as the main component to be monitored.

1.4   Main Objective of the Test Bed

In order to maintain products as competitive as possible, the detection of flaws or
maintenance needs has to be performed with a cost and time effective approach,
keeping the cost of the final product as low as possible. This includes the early
detection of machine errors. In this way, the aim of the test bed is to monitor the
performance of the CMM through the real time monitoring of the air bearings system, particularly the air flow and air pressure parameters. From the knowledge of
the technicians and operators of these machines, these parameters affect its operation. From this premise, the versatile plug-and-play platform for remote predictive
maintenance was deployed.

1.5   Needs and Motivation

In general terms, metrology is a function in the supply chain that, even though it
generates the largest amount of data, has traditionally been supported by poor infrastructure for real-time access to information. This situation has direct consequences
on the maintenance area for metrology equipment located in factories.

In the case of TRIMEK, for example, the measuring equipment provided to customers does not incorporate sensors to monitor operational parameters, in order to keep the cost of the products low and competitive, and also because this information does not provide any added-value to the clients for their purposes in buying and using TRIMEK's products. In addition, TRIMEK do not have access to the results of machines' performance once they are located in clients' facilities. As a consequence, tests need to be carried out in TRIMEK's factory and/or on-site (client's side) for gaining a certain level of insight into the possible flaws that might occur during the standard operation of the equipment or even the state of the machine in a particular moment.

In this way, the maintenance process for Bridge-type systems is carried out on site by the client and is scheduled following a theoretic approach based on a tentative number of calibrations per week/month/year. If the machine breaks down, it is necessary to check on site the reason for the malfunctioning, and in that case, a team of 1–2 people are dispatched from TRIMEK to the client's location in order to evaluate the problem and proceed with the required corrective actions. This means that, in practice, maintenance operations take place when the machine breaks down or the measurements are not accurate.

However, the wide variety of end-users results in high levels of customization which makes the number of tests and information that TRIMEK needs to collect and manage extremely large. TRIMEK is using preventive maintenance techniques based on calibration of machinery on periodic dates by the client. Calibration needs to be performed every 1 or 2 years. Verification needs to be done periodically with artefacts (tetrahedrons). This element is measured in first place with a high precision machine to know the real dimensions. Then, this tetrahedron is measured with the Bridge Machine to verify its performance (tolerances of the machine). Each client decides the maximum error margin accepted for their production. The verification is done by the M3 software and the results are stored in it, being accessible to the client/experts (Log files). As mentioned, this verification should be done daily, but clients rarely or never do it.

As manufacturing processes require high accuracy, adjusting the maintenance and calibration frequency after evaluating the results and performance or detecting failures is not recommended to the client. Therefore, the frequency or time when an equipment needs maintenance is not accurately defined. If a machine starts to measure inaccurately, has unusual or the results are not the expected ones, the probable solution could be the need of maintenance. In this case, until TRIMEK's customer warn TRIMEK about this pattern, TRIMEK could not solve it or send the maintenance team; and sometimes it is too late to fix the problem because the client usually waits too long instead of calling TRIMEK when a failure is detected.

All these factors are translated into economic losses to the competitiveness of the manufacturing company as well as deterioration of metrology solutions image and concept. The early detection of errors could reveal that the potential need for maintenance becomes imperative for the equipment manufacturer as well as for the owner in an early stage. Thus, making a shift from proactive and reactive maintenance strategies to predictive ones, could enable a better planning and scheduling of maintenance activities based on TRIMEK and client's availability, enriching the services provided by TRIMEK to its customer, increasing the useful life of the equipment and strengthening the position of the equipment manufacturing company in the market.

1.6 Expected Contribution of SERENA

Maintenance is in a time of transition to predictive concepts, or, with other words, using sensors to collect real-time operating data and artificial intelligence or machine learning to analyze the data to be able to predict when a piece of equipment needs maintenance, thereby reducing unplanned downtime and breakdowns, increasing uptime and maximizing utilization. However, this shift is beyond challenging and requires several factors to be addressed, investigated and analysed; taking into consideration the particularities of the machine, the process and the factory.

In the quality control, in-line inspections and dimensional metrology area, by implementing condition monitoring for measurement machines is expected to improve the maintenance decision-making with real, accurate and valuable data from the machine, reduce the risk of unexpected breakdowns, develop better maintenance support means for operators and technicians and improve customer experience; all within a cloud-based platform and security standards. The desire is to slightly start to introduce proactive and predictive maintenance approaches that help to increase machine availability, extend equipment lifetime and have a faster response time.

In the case of TRIMEK the aspiration is to enhance their products with advanced predictive maintenance solutions driven in SERENA project; extend its services thanks to a versatile and remote platform for predictive maintenance applicable for metrology equipment. Therefore, TRIMEK, working along with other companies in the SERENA consortium working group, focuses its activities in developing, installing and configuring the necessary devices, services and applications that will support the correct monitoring and maintenance structure concept of CMMs in the metrology factory-laboratory.

The expected outcome is mainly related to a real-time monitoring system, accessible from a cloud-based dashboard, that allows the technical teams, on one hand, to be able to remotely monitor such equipment, detect malfunctions and to significantly reduce the reaction and intervention time for maintenance activities, all this based and supported on the development and application of technologies (cloud, AR, AI-Analytics, condition monitoring) that are crucial to increase visibility of the machine condition and its failure detection, integrated with effective mechanisms to coordinate maintenance interventions on customers facilities, and also linked with knowledge and expertise from machine designer and builder for guiding technicians in problem troubleshooting and most common maintenance activities regarding CMMs.

Extrapolating this solution, through this platform TRIMEK can be be able to get and cope with the vast information produced by the equipment in the different locations and factories and add maintenance features within its own platform M3, to allow the detection of potential failures and the synchronization of maintenance with production and logistic activities, reducing the response time, improving the lifetime of the products and relations with customers.

2 SERENA System in TRIMEK Use Case

Following paragraphs describe the architecture developed for the use case, and the hardware and software solutions implemented within it, classifying the developments into the four technological areas of SERENA approach.

2.1 Remote Factory Condition and Control

A schematic representation of the test bed is presented in the following figure. The description of the modules and their correlation are presented below in Fig. 5.

As a first step, is the sensorization of the machine. To monitor the state of the air bearings, analog sensors were installed to monitor air consumption and pressure on each bearing (XYZ) (see Fig. 6). There are two X axes, which means that a total of eight sensors were installed on the machine, one in each axis. Specifically, four flow sensor and four pressure sensor with analog output that transmit current signals.

As second, a Databox was designed, developed and installed at the factory floor in the AIC TRIMEK's laboratory (see Fig. 7). The databox acts as a mediator between existing CMM, its components, M3 software, etc., the newly added data sources and sensors, and the components of the SERENA system. It consists in multiple

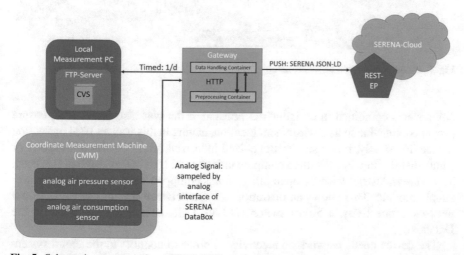

Fig. 5 Schematic representation of TRIMEK's testbed

Fig. 6 Flow and pressure sensors installed on the CMM

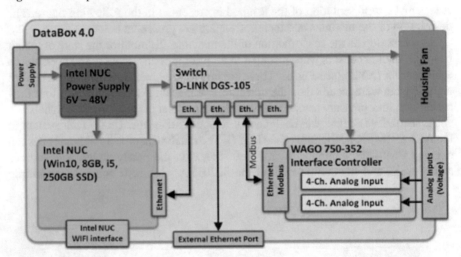

Fig. 7 Databox prototype hardware setup

components combined in an industrial housing; a fan was also included to ensure proper cooling due to its environment location, having in this way an IP24 protection degree. Physically, it is a small, ruggedized industrial field PC with a Windows or Linux distribution. For the main compute unit an Intel NUC (Type 8i7BEH) is used. To ensure industrial friendly operability, it was equipped with an additional power supply module. This allows an operation voltage between 6 V and 48V. To ensure network connectivity, a 5-port switch (D-Link DGS-105) is also included in the Databox.

The device needs network connectivity in order to connect to the cloud system via company network and the sensor interface modules. In this way, the databox

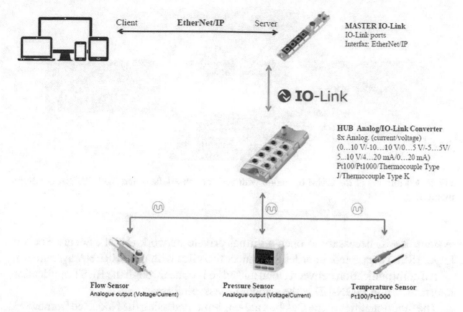

Fig. 8 Air monitoring diagram

was equipped for allowing the creation of an internal network between internal components (WAGO interface module) and additional external components (sensors). WAGO was implemented to permit safe, maintenance-free connections for a flexible signal wiring application, to perform signal transmission and distribution in the control and field levels and connect electronics to electrical systems at the control level. This interface controller was equipped with two analogue voltage input modules (WAGO 750-459). The inputs coming from the 8 sensors installed on each of the CMM axes (X, Y, Z) for monitoring the air flow and air pressure subsystem for the air bearings are current signals, so that adaptation to voltage ones was required.

For reference, the complete air system monitoring components diagram is shown in the figure (see Fig. 8).

It should be noted that the values obtained through this monitoring procedure are crosschecked against the calibration process performed with verification artefacts, in order to confirm this is the correct subsystem to monitor in relation to its influence on the quality of measurements of the machine. This is represented in the next figure (see Fig. 9).

To support also external networks, the Databox was equipped with an additional network adapter and port. In this sense, a DELL Edge gateway of the 3000 series was used to aggregate, secure, analyse and relay the data from the edge sensors, bringing both legacy systems and the sensors to the internet to obtain real-time insights of the system studied in TRIMEK's pilot. The gateway was configured to directly access and integrate the company network of TRIMEK and allow access to the SERENA cloud

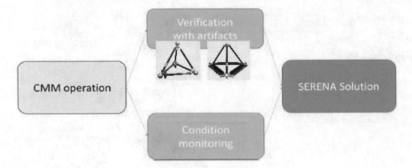

Fig. 9 Correlation of the CMM operation with verification artefacts and the SERENA condition monitoring

system; it was necessary to open a virtual private network (VPN), Secure Sockets Layer (SSL) encrypted, in order to connect the pilot with the SERENA system.

All communication between cloud and edge is done through the REST application interface, while JSON-LD is the standard message format.

The functionalities of the Databox are implemented using the NodeRed framework (see Fig. 10), based on a graphical block design and independent nodes with all functionalities needed to read and handle the incoming and outgoing data.

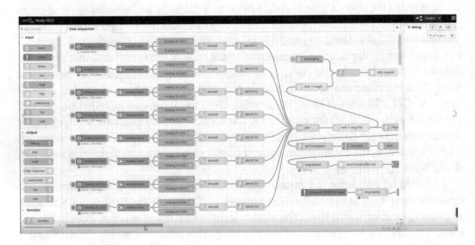

Fig. 10 Data acquisition with node red flow

The software modules are wrapped in Docker containers and stored in a local SERENA specific image repository. The Databox holds a local running data acquisition container that once acquired passes the data on to the pre-processing container; that extracts smart data from the raw data by filtering, smoothing or other. The correct timestamping of the data and a proper synchronisation of all component clocks is ensured.

In addition, MIMOSA is used to integrate condition-based maintenance information together and provide the place where to store this information (metadata, all the asset registry, condition, maintenance, reliability info. collected by the gateway).

Visualization service and dashboard:

The services included in the SERENA dashboard allows the visualization of Trimek's facilities and their production systems as a tree view control, alongside the machinery with its real-time status information, and the information about the installed sensors (see Fig. 11).

Additionally there is the option to access to other SERENA services as the scheduler.

An Interface layer with some methods has been implemented, to manage the information coming from the SERENA platform in the visualization service and vice-versa (see Fig. 12).

Its main features are:

- Visualise the real-time position on the 3D model of the machinery, in order to enable a remote monitoring of the physical behaviour observed.
- Display in real-time the status of the component (warnings, errors etc.) highlighting the involved part of the machinery with different colours, to immediately capture the operator's attention, and provide an intuitive indication of the main information to check.

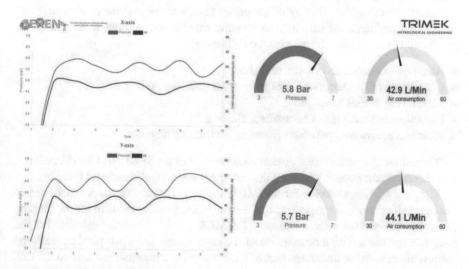

Fig. 11 TRIMEK use case data visualization

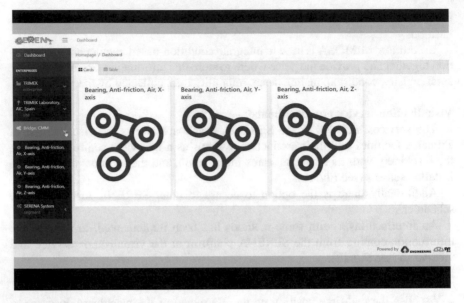

Fig. 12 SERENA dashboard: sensors description

2.2 AI Condition-Based Maintenance and Planning Techniques

For TRIMEK, it is composed by a data analytics service to accurately provide insights on correct/incorrect functioning of the equipment, and the service of planning and scheduling maintenance activities in specific timeframes without interrupting the production process plan. It includes the following:

- Analysis of historical data and real-time
- Correlation of measurement data with event data, e.g., failures
- Monitoring of real-time data
- Detection and analysis of anomalies, alerting
- Maintenance aware operations planning and scheduling

The solutions operate in the docker environment of the SERENA Cloud Platform, as the central component where all the outcomes are consolidated and integrated.

The data analytics service for TRIMEK's pilot is focused on an anomaly detection capability based on repetitions of events related to the measurements coming from the air bearings system sensors. TRIMEK provided various datasets for the analytics modules, with a description of the limit values and the typical operational values; along with the understanding of the reasons for a possible failure and the link with the originally collected data, as it is of paramount importance. The aim is to automatically detect when the air flow or air pressure are out of normal. The solution includes graphs visualization showing the cumulative anomalies, the mean values and the standard deviation in a day/week/month period, with different analytics and Key Performance Indicators (KPIs) to assess the data, like counting of anomalies, frequency of anomalies, alarm/notification when the value is out of threshold or when there are many continuous days presenting odd or wrong behaviour.

The scheduler, on the other hand, is a tool built as a web application, that supports the connection to the central database containing the data required for planning/scheduling of the maintenance activities. Its interface is used for visualizing multiple information and results and interact with Machines, Scheduling, Tasks and Resources UIs.

As the final visual result is offered a Gantt chart for visualizing the result of the scheduling algorithm, with the different maintenance tasks assigned to the different resources in a time slot. In addition, another table is visualized providing short information about the duration of a task and the maintenance tasks scheduled. The company information is received by the SERENA Dashboard, then the Scheduling Tool automatically retrieve TRIMEK's info and workflows from the AR Workflows module, mapped and stored as resource and tasks.

2.3 AR-Based Technologies for Remote Assistance and Human Operator Support

The system for AR-based step-by-step instructions to support service technicians on-site is a web application that can be used on daily basis on computers, tablets and phones to perform maintenance tasks, whether they are scheduled maintenance cases or it is a training for new personnel. The tasks can even be performed without internet connection having the manuals stored locally on the device. For this purpose, the application provides an instruction authoring interface for editing instruction sets to a database; to create multi-media instructions, correlating to experience levels and maintenance operations as well as display devices. A TRIMEK administrator or a maintenance expert can insert text, image, audio instruction and stored them. This was the case for the demonstrator: A set of videos and texts were uploaded regarding air filters unit maintenance. The system was developed by OCULAVIS and the output is the outcome of the user interaction with this interface constituting one or more maintenance instruction sets. In this context, as soon as a maintenance activity has been assigned to an operator by the scheduling tool, the operator support services are triggered.

2.4 Cloud-Based Platform for Versatile Remote Diagnostics

For all the services developed for TRIMEK, the related cloud storage are implemented in a dedicated SERENA tenant cloud on DELL's Infinite testbed, as was

Fig. 13 SERENA dashboard

mentioned before. All service can be accessed through the SERENA dashboard (see Fig. 13).

3   Results

The demonstrator of the test bed consists on evaluate the following scenario: each machine axes (X, Y, Z) is equipped with air pressure and air flow sensors to monitor one of the most critical elements of these machines, the air bearings system, as problems in these components induce errors in the measurements and lack of repeatability on the outputs. The gathered data coming from the sensors is transferred to the SERENA cloud-based platform in order to have a control monitoring system available on a dashboard. Then, a data analytics process is performed on the data in order to detect anomalies in the air flow or pressure (real or simulated) based on thresholds. With this information, the operator has valuable information about the state of the machine, and if necessary, a maintenance visit is scheduled with the specific planning service. Finally, through the operator support system the maintenance instructions are given to the operator to perform the required work.

SERENA solution was validated in TRIMEK's laboratory at the Automotive Intelligence Center where the required HW and SW are available. Although, final benefits are thought for a situation where the machine is already sold to a customer and it's placed in the client facility or factory. The impact of SERENA based on the experimentation results was assessed by personnel with expertise in the use of the machine, its maintenance, the client's involvement, operators and operations coordinators, through the seek of the following objectives:

- Improve the portfolio of services provided with the sale of TRIMEK's products, strengthening its market position and increasing its competitiveness

- Increase customer satisfaction with the overall functioning of the machine, its state monitoring and new maintenance approach
- Facilitate the knowledge of TRIMEK's different types of client machine's state and maintenance related data management
- Assess a shift from proactive and reactive maintenance strategies to predictive, at least in a partial manner
- Reduce the soft and hard number of machine breakdowns per year
- Reduce the cost associated to repairing and maintenance activities
- Extend the CMM machine lifecycle
- Improve the accurateness of the maintenance personnel sent to customers for problem solution and maintenance activities

On the other hand, after the deployment and configuration of the solution within the test bed, the features to be tested were:

- Access to the SERENA dashboard
- Remote air flow and pressure sensor data collection, data acquisition and flowing (remote condition monitoring)
- Data analytics service for anomaly detection based on sensor data
- Planning and scheduling service for maintenance activities
- Operator support system for maintenance activities related instructions regarding air bearing component

In this sense, four major experiments were held and are described in the following paragraphs.

Experiment 1. Proper visualization of air bearings flow and pressure sensor data collected from the CMM

The objective is to facilitate the knowledge of TRIMEK's different types of client machine's state by evaluating the provided databox functioning in TRIMEK's laboratory along with the air pressure and air flow sensors installed. The projection is to verify the possibility of increasing customer satisfaction with a remote monitoring of an accuracy related parameter, reducing failures and costs, and extending the lifetime of CMM.

The experiment consisted on turning on the gateway for a period of time in order to collect a set of data from the air bearings of each machine axe, and then, visualize its representation through the SERENA Databox Dashboard developed for TRIMEK; that shows voltages inputs, two for each side of machine's X axe, one for Y, and one for Z. The features to be tested were: remote air flow and pressure sensor data collection, data acquisition and flowing (remote condition monitoring), functioning of the databox, sensor wiring and connections, access to the SERENA Databox Dashboard, correct visualization and representation of sensor data. In other words, to be able to monitor the air bearing system of the CMM machine.

Two datasets of sensor data from different periods (2 weeks approx./each one) were collected. In both times, the gateway functioned as it was supposed to and the data was correctly collected and visualized through the corresponding dashboard, and also stored in SERENA's cloud. In this way, TRIMEK was able to monitor in real-time the air flow and air pressure signals of each of the machine axes.

The dashboard for visualization is illustrated in the following figures (see Figs. 14 and 15).

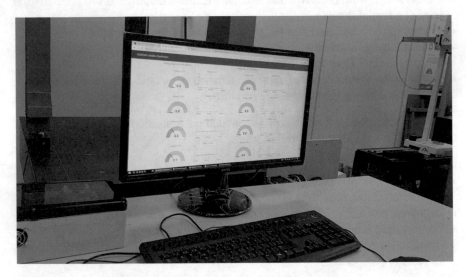

Fig. 14 TRIMEK use case data visualization

Fig. 15 Visualization dashboard

Experiment 2. Have an analysis on the collected air bearing system data through an anomaly detection service

The objective is to have a deep knowledge of the state of the machine based on the air bearings subsystem by using the analytics service developed for TRIMEK which is focused on anomaly detection of sensor data out of threshold or with unusual frequency of occurrence. The projection is based on this, being able to reduce the occurrence of failures by detecting the issue in an early stage, extending the life-cycle of the machine, assuming a more proactive and fully preventive maintenance strategy, reducing at the same time the maintenance costs.

As the system is not expected to fail (the pressure and air flow are not expected to overpass the defined limits), the air supply system was manipulated to create artificial losses in the air system unit so that the sensors can sense these wrong values and the data analytics tool can perform an analysis on the data, providing some visual insights of the system status, showing the situation in real-time. The system is expected to show graphs for visualization of the behaviour of the air flow and pressure values and measure the counting and frequency of anomalies, plus showing a notification when the values are passing the threshold.

A group of KPIs were defined for, particularly, being able to measure or analyse the results graphs provided by the sensor data collected and processed. They are the following:

Counting of anomalies: Number of anomalies per day/week, Frequency of anomalies, Cumulative number of anomalies over a week, Distance between anomalies (in days) Notification: When the values are out of the ±20% of threshold, When there are 3 continues days of high values, When the number of anomalies per day is higher than 15 to notify odd behaviour and higher than 20 to notify a critical issue.

In this sense, the first colour notification will inform the operator that something might be happening and that he should in the coming days review the data and graphs to make an assessment of it, and a second colour notification will inform that is something more serious that need an assessment sooner as possible.

These resulting graphs and analytics are shown in the next figures (see Figs. 16, 17 and 18), particularly for the Y axis. These graphs are intended to give a quick deep knowledge to the operator and serve as a base to make further assessments and decisions on maintenance needs.

However, the service acquire more significance when the datasets are referred to longer periods (years).

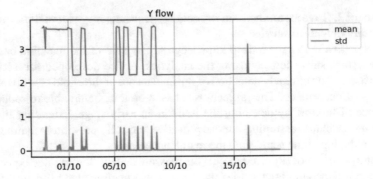

Fig. 16 Graphs results from data analytic service: mean and standard deviation value over time

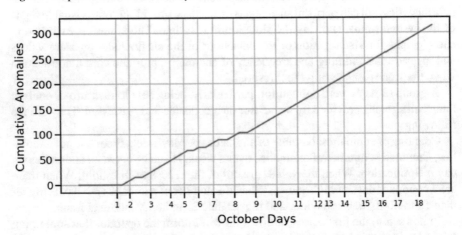

Fig. 17 Graphs results from data analytic service: cumulative anomalies over time

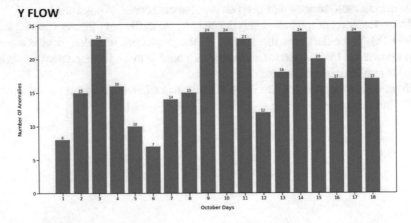

Fig. 18 Graphs results from data analytic service: number of anomalies per day over time

Experiment 3. Schedule a maintenance visit within the adequate conditions

The objective is to evaluate the scheduling tool by scheduling a maintenance visit within the adequate conditions taking in consideration the disturbance and availability in the company and clients. Also, validate all the sections in the scheduler application, the inputs required from the user side, and the proper visualization and understanding, besides verifying the proper retrieval of the workflows from the operator support systems. The projection is to improve, by the use of this tool, the accurateness of the maintenance personnel sent to customers for problem solving and maintenance activities, the planification of the overall procedure and the automation at some level of it, having a more organized maintenance protocol

The experiment consisted on validating the scheduling tool and its interfaces based on the interaction of the user/operator with the tool, obtaining as result a visit scheduled on a date and time where it causes the least possible disturbance for the company, incrementing the operator satisfaction, the availability of the machine, and reducing the time spent on the whole maintenance protocol, from the noticing of a failure, to the scheduling process in the factory and with the customer, and the maintenance execution.

In these sense, based on a simulated alarm or notification of that values from air flow and pressure were odd or bad, the operator after assessing previous service result graphs and KPIs, accessed the scheduling tool through the SERENA dashboard, introduced the required inputs and successfully obtained a scheduled fictional visit for the application of the maintenance task related to air filters maintenance (assuming this is the issue solution, for the demonstrator). TRIMEK information was retrieved, available resources were added (technician), maintenance task was retrieved from the operator support platform and linked with the available technician. After this, it was received the proper scheduling of the activity, presented in the Gantt chart (see Fig. 19).

This scheduler, during a regular use, will help reduce the time needed for the machine to be operable again, thus, increasing the availability of the machine and the turnover as well as decreasing the costs associated to production breakdowns and the costs associated to personnel as the workload can be easily distributed.

Experiment 4. Perform the air filter unit maintenance following the operator support workflow

The projection is to improve the complete maintenance protocol procedure for a maintenance visit with the use of the operator support workflow tool system deployed for TRIMEK. The objective is to perform a maintenance task following the instructions uploaded to the tool regarding the air bearings component selected for the project study; accessing the tool, navigating through it and obtaining the knowledge by interacting with it.

The experiment consists on completing a maintenance activity (see Figs. 20 and 21) with the expected outcome in a shorter time than without the use of the operator support system and reducing the need of 2 or more personnel for the successful achievement of the task.

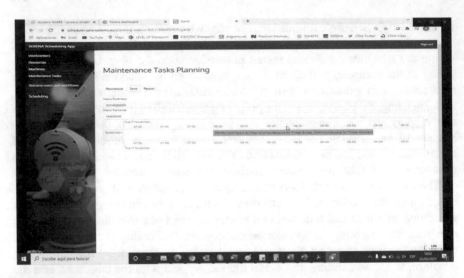

Fig. 19 Scheduler user interface

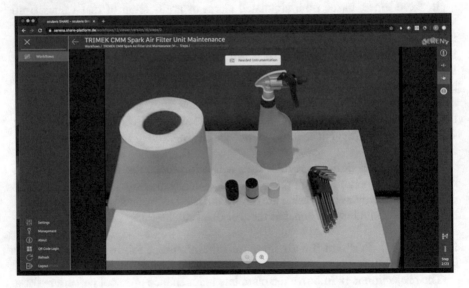

Fig. 20 Video instruction for maintenance task

Likewise, with this service is expected to increment the operator satisfaction during the process in 2 points in a scale from 1 to 10, reduce the personnel needed for the process to one and the time required, improving in this sense the availability of the machine. In this way, based on a simulated scheduled visit for the maintenance of the air system, the operator will access the operator support system interface through laptop, table or phone and follow a maintenance workflow regarding the replacement of the air filtering unit in a CMM, following step by step the instructions.

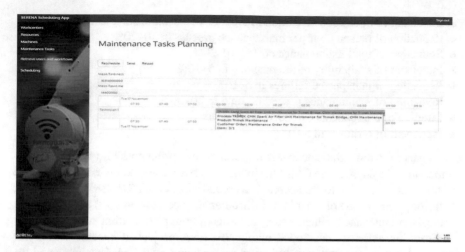

Fig. 21 Scheduler tool retrieving workflows from operator support platform

The result was that the operator successfully accessed and moved through the tool, and executed the maintenance task related to SPARK CMM air filter unit maintenance, with a great understanding on the instructions, text and videos uploaded to the workflow in the platform.

Finally, by assessing the results from validation period it was concluded that they are correlated to conform the general SERENA solution. In this sense, being able to monitor the machine in real time, and remotely, helps to improve production breakdowns frequency, calibration control and early detection of failures. This reality definitely affects the maintenance costs by tackling the machine issues on time, avoiding the appearance of a major damage. The monitoring is complemented with the data analytics, visualization and the anomalies detection, which provides to the operator a deep knowledge about the state of the air bearings system and the machine, plus more specific information to make further decisions about if it is needed a maintenance visit and the level of urgency.

The scheduler tool represents a way for better planning considering clients and company availability, production schedule, workload, etc. Besides the manner of presenting the information and pending tasks through the Gantt chart. By its use, along with the operator support workflow system, the need of technicians per visit is reduced because the technical needs of the visit are predetermined and knowing which components have failed there is just need for a technician specialized in those specific components and technologies; counting with the fact that it will be more prepared maintenance personnel with the use of the maintenance related instructions presented on the computer, tablet or phone.

In summary, all the tools and services functioning, correlation, assessment and validation from TRIMEK's perspective as end user can be translated into the following:

- Increment of operator satisfaction in 2 points in a scale from 1 to 10
- Reduction of person hour per maintenance process in 10–20%
- Reduction of total maintenance cost in 10–20%
- Increment of availability of the machine in 5–10%
- Reduction of unexpected failures in 5–10%
- Reduction of maintenance procedure time in 20–30%
- Lifetime of components extended in 0.5 years
- Increment of turnover in 5–10%

In general words, the outcome is useful in terms of the possibility of monitor an important subsystem in the CMM that permits to have a deep knowledge on the state of the machine related to the accuracy, a crucial aspect for TRIMEK and its clients; that along with the use of other tools (scheduler and operator support) allows to better solve the maintenance related issues, in terms of both planification and execution.

From the activities developed within SERENA project, it became clearer the relation between the accuracy and good performance of a CMM machine and the good functioning of the air bearings system; it was proved or confirmed by TRIMEK's machine operators and personnel that air flow or pressure inputs and values out of the defined threshold directly affects the machine accuracy. This outcome was possible from the correlation made with the datasets collected from the sensors installed in the machine axes and the use of the tetrahedron artefact to make a verification of the accuracy of the machine, thanks in first instance, to the remote real-time monitoring system deployed for TRIMEK's pilot. This has helped to reflect the importance of cost and time effective maintenance approach, and the need of being able to monitor critical parameters of the machine, both for TRIMEK as company and for the client's perspective.

Another outcome learned throughout the project development is in relation to the AI maintenance-based techniques, as it is required multiple large datasets besides the requirement of having failure data in order to develop accurate algorithms; and based on that CMM machine are usually very stable this makes difficult to develop algorithms for a fully predictive maintenance approach in metrology sector, at least with a short period of time for collection and assessment.

In other sense, it became visible that the operator support system is a valuable tool for TRIMEK's personnel, mainly the operators (and new operators) as an intuitive and interacting guide for performing maintenance task that can be more exploited adding more workflows to other maintenance activities apart from the air bearings system. Additionally, a more customized scheduler could also represent a useful tool for the daily use with customers.

By having the real time monitoring of such important system as the air bearings, the maintenance approach can be more predictive and preventive; more practical and beneficial in relation to the personnel activities. Besides the maintenance operator support really helps to understand the tasks needed to be performed, so the satisfaction in a general perspective is improved. SERENA platform and maintenance operator support represent a user-friendly and effective way to perform training to new personnel, besides it definitely improves the level and value of detailed information by having it on videos, audios, etc., allowing the company to organize such processes in a more effective way. As all new software and hardware, it is needed to learn how to used and install it, but TRIMEK's personnel is used to manage digital tools

In general words, SERENA, represents a maintenance approach with a lot of potential and value from TRIMEK's perspective. The different systems and services developed for TRIMEK and concentrated in SERENA platform, working together as a "chain of services" have a positive influence on the occurrence and early detection of breakdowns, as well as in the time for solving the issues stopping the machine availability; by having the opportunity to monitor the state of an important subsystem for the machine accuracy assurance and visualize this valuable data, allowing the operator to assess the situation, with real-time and historical data, for making the best decision on the maintenance actions that need the machine in an early stage. This, for sure, impacts the lifecycle of the machine.

4 Conclusion

Metrology engineering industry was one of the main areas where to implement SERENA project for its validation. The work was based on the study and development of four technical areas. These are:

- Remote factory condition monitoring and control
- AI condition-based maintenance and planning techniques
- AR-based technologies for remote assistance and human operator support
- Cloud-based platform for versatile remote diagnostics.

For TRIMEK test bed, each area was developed with the adaptation to the particular technical requirements and instances present in the laboratory of TRIMEK with the machine. Different levels of advance were obtained in each of these. The most beneficial from TRIMEK's perspective as end user were related to the implementation of the remote cloud-based condition monitoring system and the platform for operator support in maintenance tasks. Nevertheless, it was validated that SERENA approach represents, indeed, a versatile plug-and-play platform regarding industrial machines maintenance, offering diverse services centralised in a single general platform.

An important outcome that became clearer after research and development activities in the process of complying SERENA project objectives, was that air bearings system of the CMM are one of the most important components of these type of machines; extremely related with machine accuracy and performance. Based on this fact, the sensorization of the machine working along with a mechanism for data collection integrated on a network cloud-based, that permits to remotely monitor air supply and pressure parameters and visualise them on a dashboard, with a procedure for data analysis, represents an important advance that can be exploited and improved in future. The approach has the potential of helping in the migration to better maintenance methods, more practical, preventive and even predictive. As it was mentioned before in this chapter, in the case of TRIMEK, when machines are sold to customers there is no available form to remotely monitor the state of the machine, implying that currently maintenance is only corrective and preventive, with the consequences this has for the company and for the customers.

On the other hand, despite the training of data sets for the development of predictive algorithms was challenging, the proposed analytic methodology for detecting anomalies and abnormal events is promising. In a first approach the difficulties were related to the large amount of data needed, specially failure data, and the fact that CMMs are very stables machines. However, it is required much longer periods of data collection for the development of more elaborated algorithms that represents accurately the behaviour of the machine subsystem under study [3, 4].

In general, the whole concept applied in metrology machines can be taking into consideration by other companies in the sector selling metrology equipment, in order to improve their maintenance practices when machines are sold to customers and installed at their factories. Additionally, two other tools are included for helping to achieve this: the scheduler, for a better organization of the maintenance interventions with improved decision time, and the didactic operator support, to guide maintenance personnel in the machines interventions.

As a joint, the overall solution can definitely enrich TRIMEK's service provided to customers which was the end goal of the company in participating as a pilot line in SERENA project.

Acknowledgements This research has been partially funded by the European project "SERENA—VerSatilE plug-and-play platform enabling REmote predictive mainteNAnce" (Grant Agreement: 767561).

References

1. S. Panicucci, N. Nikolakis, T. Cerquitelli, F. Ventura, S. Proto, E. Macii, S. Makris, D. Bowden, P. Becker, N. O'Mahony, L. Morabito, C. Napione, A. Marguglio, G. Coppo, S. Andolina, A cloud-to-edge approach to support predictive analytics in robotics industry. Electronics **9**(3) (2020). https://www.mdpi.com/2079-9292/9/3/492
2. T. Cerquitelli, D. Bowden, A. Marguglio, L. Morabito, C. Napione, S. Panicucci, N. Nikolakis, S. Makris, G. Coppo, S. Andolina, A. Macii, E. Macii, N. O'Mahony, P. Becker, S. Jung, A fog computing approach for predictive maintenance, in advanced information systems engineering workshops - caise, international workshops, Rome, Italy, 3–7 June 2019, *Proceedings, Lecture Notes in Business Information Processing*, vol. 349, ed. by H.A. Proper, J. Stirna (Springer, 2019). Lect. Notes Bus. Inf. Process. **349**, 139–147 (2019). https://doi.org/10.1007/978-3-030-20948-3_13
3. T. Cerquitelli, D.J. Pagliari, A. Calimera, L. Bottaccioli, E. Patti, A. Acquaviva, M. Poncino, Manufacturing as a data-driven practice: Methodologies, technologies, and tools. Proc. IEEE **109**(4), 399 (2021). https://doi.org/10.1109/JPROC.2021.3056006

4. D. Apiletti, C. Barberis, T. Cerquitelli, A. Macii, E. Macii, M. Poncino, F. Ventura, istep, an integrated self-tuning engine for predictive maintenance in industry 4.0, in *IEEE International Conference on Parallel and Distributed Processing with Applications, Ubiquitous Computing and Communications, Big Data & Cloud Computing, Social Computing and Networking, Sustainable Computing and Communications, ISPA/IUCC/BDCloud/SocialCom/SustainCom 2018, Melbourne, Australia, December 11-13, 2018*, ed. by J. Chen, L.T. Yang (IEEE, 2018), pp. 924–931. https://doi.org/10.1109/BDCloud.2018.00136

Estimating Remaining Useful Life: A Data-Driven Methodology for the White Goods Industry

Pierluigi Petrali, Tania Cerquitelli, Paolo Bethaz, Lia Morra, Nikolaos Nikolakis, Claudia De Vizia, and Enrico Macii

Abstract A Predictive Maintenance strategy for a complex machine requires a sophisticated and non-trivial analytical stage to provide accurate and trusted predictions. It must be planned and carried out carefully to maximise the information extracted from available data. The SERENA project provided an excellent methodological framework and a solid technical and software foundation to deliver a robust and applicable Predictive Maintenance solution for the White Goods industry. The proposed data-driven methodology was applied on a real use case, that is, estimating the degradation trend of industrial foaming machines and predicting their remaining useful life. The models were built based on historical data and are applied in real-time adjourning their predictions every time new data are collected. The results are promising and highlight how the proposed methodology can be used to achieve a fairly accurate estimate of machinery degradation and plan maintenance interventions accordingly, with significant savings in terms of costs and time.

P. Petrali (✉)
Whirlpool EMEA, Biandronno, VA, Italy
e-mail: pierluigi_petrali@whirlpool.com

T. Cerquitelli · P. Bethaz · L. Morra · C. De Vizia
Department of Control and Computer Engineering, Politecnico di Torino, Turin, Italy
e-mail: tania.cerquitelli@polito.it

P. Bethaz
e-mail: paolo.bethaz@polito.it

L. Morra
e-mail: lia.morra@polito.it

C. De Vizia
e-mail: claudia.devizia@polito.it

N. Nikolakis
Laboratory for Manufacturing Systems and Automation, University of Patras, Patras, Greece
e-mail: nikolakis@lms.mech.upatras.gr

E. Macii
Interuniversity Department of Regional and Urban Studies and Planning, Politecnico di Torino, Turin, Italy
e-mail: enrico.Macii@polito.it

© Springer Nature Singapore Pte Ltd. 2021
T. Cerquitelli et al. (eds.), *Predictive Maintenance in Smart Factories*,
Information Fusion and Data Science,
https://doi.org/10.1007/978-981-16-2940-2_7

149

1 Introduction

Nowadays, thanks to the technological evolution from which many industrial scenarios can benefit, Industry 4.0-enabled manufacturing sites are increasingly more frequent. In these scenarios, the challenge is to monitor the entire production chain through sensors that can constantly collect a large amount of data. This data, processed by means of scalable data analytics architectures, can be leveraged to optimise the machinery maintenance process. In fact, being able to forecast when a machine will fail is a key objective in an industrial context, which can lead to great cost savings thanks to a predictive maintenance intervention, avoiding forced interruptions in production activities.

In this chapter, we focus on an industrial use case which involves the monitoring of industrial equipment at Whirlpool Corporation, a leader in the white goods industry. In particular, data were collected from industrial foaming machines, with the aim of predicting their level of degradation and then intervening promptly with a maintenance service. In this scenario, a lot of data is collected from different sensors over a fairly long period of time. Some signals represent a fully functioning machine condition, others have been collected in the proximity of a failure and therefore represent critical production issues. However, detecting and understanding the physical events that lead to the failure is not a trivial task. Therefore, we here propose a scalable predictive maintenance service capable of estimating machinery degradation behaviour over time. After defining a degradation profile by analysing historical data, the service exploits this knowledge to offer online predictions on new data collected in real-time. Furthermore, to offer more interpretable and actionable results to the industrial stakeholders, besides forecasting the relative state of degradation of the machine (in percentage), the service also estimates the Remaining Useful Life (RUL) expressed as the number of cycles the machine is still capable of completing. This valuable information can be directly exploited to better schedule the production cycles up to the next maintenance intervention.

The rest of this chapter is organised as follows. Section 2 presents in detail the use case of interest, describing the industrial process and maintenance operations required, while Sect. 3 describes previous state-of-the-art works in literature related to the proposed scenario. Section 4 describes the strategy implemented to estimate machine degradation and RUL, as well as how the Whirlpool use case is integrated into the Serena architecture. Experimental settings and results are given in Sect. 5. Finally, Sects. 6 and 7 highlight important aspects that emerged during the experiments, drawing conclusions and presenting possible future improvements.

2 The White Goods Industry: Use Case Description

Whirlpool Corporation is the world's leading manufacturer and marketer of major home appliances, with annual sales of more than US\$18 billion, with more than 73,000 employees and more than 70 manufacturing and technology research centres around the world. In EMEA Region Whirlpool is present with its headquarters based in Italy and 15 production sites in Italy, Poland, Great Britain, Slovakia, Russia and Turkey. The average factory size is able to produce from 500.000 to 1.000.000 pieces per year, serving a higher number of brands, models, and countries. Specifically, the SERENA use case is based on the Italian manufacturing sites where built-in refrigerators are produced.

2.1 Industrial Process

A refrigerator is insulated by means of a rigid polyurethane foam, characterised by low density, high mechanical strength and optimal thermal insulation properties. Among the different production processes involved in the manufacturing of refrigerators, Polyurethane Foaming is the core step, and one of the most complex in the White Goods industry. Polyurethane basic components are injected in liquid form inside the cavity existing between the internal plastic liners and the outside containment panels (made of various material such as steel, plastic, cardboard, etc.), and the entire polymerisation phase happens in there.

Typically, the process to achieve this comprises four main steps: *storage, mixing, pouring, and final assembly*, as depicted in Fig. 1. Basic chemicals, that are Polyols, Isocyanate and Blowing Agents (BA), are stored separately in the Bulk **storage** area. In the second step, Polyols and BA (typically a hydrocarbon gas such as Cyclopentane) are premixed in a mixing station. A dosing system then prepares the right

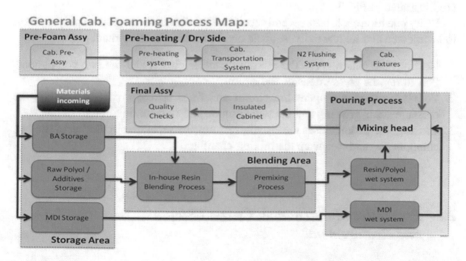

Fig. 1 The foaming process comprises four main steps: storage of the basic chemicals, mixing of the polyurethane components, pouring and final assembly. The polymerising reaction takes place in the pouring process, where the liquid mixture in injected inside a mould

Fig. 2 Schematic
representation of the mixing
head

Isocyanate (MDI)

Polyol

Self Cleaning piston

amount of components and sends them at controlled temperature to an injection system, where the cabinets are kept inside a mould while the liquid mixture is injected (**pouring** stage). Since the polymerising reaction, exothermic and expanding, can last several minutes, the mould and cabinets have to be kept under specific jigs to retain the shape and optimise cooling. A self-sustained cabinet can be sent to the final **assembly** stage to be equipped with the remaining parts of the cooling circuits, accessories, user interfaces, doors etc.

The most critical part of the entire process is the **mixing head**, a complex system that allows the two components incoming at high pressure to be mixed and injected in the refrigeration cavity. The mixing head, depicted in Fig. 2, is also characterised by a moving element called **self-cleaning piston** which has the double function of opening and closing the component circuit at high pressure, and cleaning up the mixing chamber to prevent residuals of the mixture from polymerising inside it, compromising its functionality. The self-cleaning piston and its main functionalities are illustrated in Fig. 3.

Every injection cycle lasts several seconds according to the amount of polyurethane to be injected and its reactivity. A cycle can be divided in three main phases:

1. **Resting**: the piston keeps the mixing chamber closed and the two components flow at regime pressure inside the circuit;
2. **Pouring**: the piston opens the circuit and frees up the mixing chambers; the pressure starts growing and the two components start mixing in the chamber;
3. **Closing**: the piston closes the circuit, restores the separate circulation of the two components, and cleans up the mixing chamber.

2.2 Preventive Maintenance: Current Status and Requirements

The mixing head undergoes a natural wearing due to the mechanical stress and friction of the piston, and therefore it is crucial to arrange a suitable preventive plan. In this Section, the transition from a traditional to a predictive, data-driven

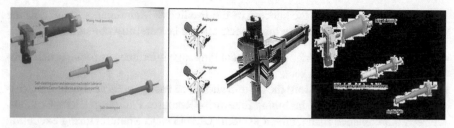

Fig. 3 Schematic representation of the mixing head assembly, including the self-cleaning piston (right). The piston's main function is to open and close the mixing chamber in the Pouring and Resting phases, respectively (left)

maintenance strategy will be addressed. Specifically, moving from the unfulfilled needs of traditional strategies, the business requirements for data-driven strategies are outlined.

In conventional preventive maintenance plans, the mixing head is substituted *after a specific number of shots* is reached. In fact, the head after a normal number of cycles is typically restored to its initial condition through some mechanical working. Even though this schedule is carefully calibrated, unexpected breakdowns can still occur before the substitution planned date, resulting in equipment downtime and higher maintenance cost. On the other hand, since the wearing process is variable and depends on many uncontrollable factors, in many cases the head is substituted in advance with respect to its actual wearing. With a predictive maintenance strategy, the availability of the machine and the efficiency of the foaming process could be increased, avoiding unnecessary substitution.

The SERENA methodology was applied to this use case with three main objectives in mind:

- **Avoid** potential **breakdowns** associated with a fixed schedule preventive maintenance strategy;
- **Maximise the operating life** of the component;
- **Reduce the total cost of maintenance**.

This strategy is supported by three key enabling technologies:

- **Remote condition monitoring and control**: the lack of sensors made particularly difficult to monitor the mixing head state. Within the use case, additional sensors, as well as the parameters to monitor, were defined based on several shots of the mixing head. These sensors complement existing data already recorded and available through the SCADA (Supervisory control and data acquisition) systems.
- **Predictive maintenance and planning techniques**: *machine learning* techniques leverage historical datasets to correlate sensor data with failures mode and classify the level of technical intervention needed.
- **Augmented Reality (AR) interfaces**: such tools, coupled with wearable devices, have the potential to improve maintenance and enable a more flexible and effective organisation of work.

Additionally, to ensure the success of the Predictive Maintenance system in the white goods industry, the following aspects should be carefully considered:

- **System usability**: user friendly and local language interfaces improve acceptability by the shop floor workers;
- **Legacy system integration**: the system should be easily integrated with the existing information systems, including Enterprise Resource Planning (ERP), Computerised maintenance management system (CMMS) and Manufacturing execution system (MES) system, in order to gather more data to enhance the prediction (e.g., status of the production), as well as to ensure the timely and seamlessly integration of the predictions in management systems (e.g., automatic order generation in CMMS).

3 Literature Review

In recent years, Industry 4.0 has revolutionized the way of working inside smart factories. The monitoring of the entire production phase and the storage of the data collected involves the use of sensors, clouds and suitable networks for the communication between the various components. This whole scenario is well illustrated in [1], where the authors describe the operation of an industrial intelligent environment. The data obtained during this process, if collected properly, can be leveraged as a key element in the decision-making process of a company. For this reason, the industrial world is now also empowered with data mining and machine learning algorithms that, thanks to their data-driven nature, can extract useful insights from the collected data, that can lead to important production decisions. Good data quality is a fundamental condition to obtain valid and exploitable results. Methods to improve data quality are discussed in [2], where authors offer a detailed explanation of how to treat the data in order to eliminate outliers.

Dalzochio et al. [3] collects a series of works carried out from 2015 to 2020, based on the use of machine learning techniques to facilitate predictive maintenance in manufacturing contexts. Motahari-Nezhad and Jafari [4] proposes a data-driven bearing performance degradation assessment method, while in [5] authors introduce a lightweight Industrial Internet of Things (IIoT) platform for identifying the symptoms of imminent machine failure through a predictive analysis. In [6], a new framework is presented, providing a scalable Big Data service able to predict alarming conditions in slowly-degrading processes characterized by cyclic procedures. Knowing how to properly manage huge amounts of data is essential in an environment where the production process is constantly monitored, generating a very high number of collected parameters. [7] assesses the adaptation of IT tools in managing industrial Big Data and predictive maintenance operations, while another application of Big Data frameworks used in a predictive maintenance context is done in [8], where a self-tuning engine for analyzinf data in real-time is presented, exploiting a distributed architecture based on Apache Kafka, Spark Streaming, MLlib, and Cassandra.

4 The SERENA Data Analytics Pipeline

In this section, we describe the methodology used to build a predictive model capable of estimating the RUL of the mixing head. After selecting a signal to be monitored, the key concept is to define an 'ideal' behaviour of that signal, in which degradation is assumed to be zero. Then, each time new data is collected and processed, it is compared with the ideal expected behaviour to obtain an estimate of the mixing head degradation. The diagram in Fig. 4 summarises the main steps included in the proposed methodology. Each of the mentioned blocks is described in detail below, after a brief description of the main data sources.

4.1 Data Sources

The SCADA system collects real time data for every injection cycle, and stores it in a SQL database. There are three main sources of information:

1. Data about the components during the injection: temperatures, pressures and mass flow of the two components measured at the head;
2. Data about the mixing head: Oil pressure of the piston and mechanical vibration recorded during the opening and closing phases;
3. Alarms recorded by the systems.

All data are timestamped to ensure that the time series can be properly reconstructed. A second source of data is provided by past recording of maintenance activities, such as occurrence of failures, time of failures, and recovery actions. This

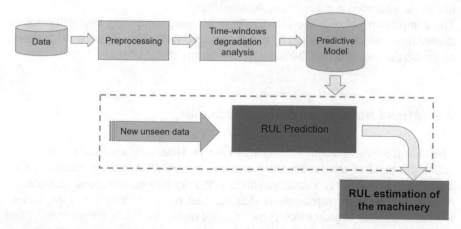

Fig. 4 Main components of the proposed data-driven Remaining Useful Life (RUL) prediction methodology

dataset is composed from several sources, including the existing preventive maintenance dataset and additional spreadsheets manually compiled by maintenance operators.

4.2 Preprocessing

First of all, in the preprocessing step, statistical features are extracted from each
signal of the production cycle. In particular, signals are split in contiguous portions
to better capture the variability of the series. Then, for each split, statistical features
able to summarise the time-series trend are computed (e.g., mean, standard deviation, kurtosis, skewness, root mean squared error, sum of absolute values, quartiles,
number of elements over the mean, absolute energy, mean absolute change). Finally,
only the most informative statistical features are selected, after evaluating the correlation between all each pair attributes and eliminating those that were most correlated
(based on the average correlation of all other pairs).

4.3 Time-Window Degradation Analysis

The proposed methodology, as previously mentioned, is based on the comparison
between time windows containing different production cycles over time. The first
is the reference window, representing the optimal condition of the machinery, in
which degradation is absent (or assumed to be). The others are moving windows
containing the new signals collected over time, and whose distribution is compared
to the reference one. The higher the difference between the two distributions, the
higher the estimated degradation of the machinery.
The comparison between the two windows involves comparing two multivariate
distributions, which is achieved by employing the Wasserstein distance [9], a state-
of-the-art distance function defined over probability distributions.

4.4 Model Building and RUL Prediction

The distance measurements calculated in the previous step are here exploited to
define the trend of the machine degradation. In particular, a linear model of the
degradation processed is defined and fitted to the distance measurements calculated
over time. The linear regression is characterised by an intercept, a slope, and a
maximum number of production cycles reached during the life of the machine. This
model will be the key point for estimating the degradation of new production cycles
(unseen data). In particular, each time measurements on new production cycles are
collected, they will constitute a new time window. As usual, the distance between

this window and the reference window is calculated. After that, it is possible to assess where this distance lies on the previously defined degradation model, thus estimating the degradation of the machine and the number of production cycles it is still able to withstand before maintenance is required. This information on the remaining days of production constitutes the RUL value of the machine.

4.5 Deployment

The predictive methodology proposed here was implemented in a Python package and integrated as a service in the SERENA cloud platform, described in Chapter "A Hybrid Cloud-to-Edge Predictive Maintenance Platform" and Chapter "Services to Facilitate Predictive Maintenance in Industry 4.0". A scheme of this edge-cloud interaction is illustrated in Fig. 5. The predictive service on the cloud is implemented as a Docker container, enabling scalability and modularity. The SERENA platform is connected to shop floor data through a data gateway run on an edge computer on the Whirlpool premises. The function of the gateway is to gather data from the database and transfer them into the cloud through a Virtual Private Network, keeping data consistency and transforming them in the preferred format for SERENA modules. A two-way TLS authentication is implemented by a Reverse Proxy Certification Authority (RPCA), that accepts or rejects the incoming requests, based on a set of security policies. The different services on the cloud communicate with each other through JSON-LD [10] messages, which format is partially built upon the MIMOSA [11] open standard for asset maintenance and, in particular, the CRIS 3.2 schema. The data analysed on the cloud generate RUL predictions, which are presented through a Dashboard.

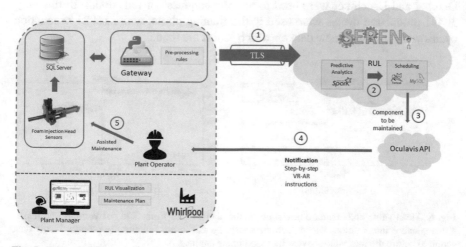

Fig. 5 Edge-cloud interaction for the predictive service

5 Results

This section shows the results obtained using the techniques described in Sect. 4, with the goal to verify how well the real-time prediction methodology can estimate the degradation of the machinery.

After consultation with the domain experts, the "Force, Self-cleaning Piston, closing" and "Force, Self-cleaning Piston, opening" were chosen as the signals to be monitored. These signals were collected during an opening/closing phase of the piston and were therefore considered good candidates from which to predict the state of degradation of the machinery. The available data was collected between October 2019 and July 2020: a total of 86960 production cycles over 10 months. During these 10 months, two known maintenance operations were carried out on the head, one in early December and one in early May. A full machine life cycle was therefore available from December 2019 to May 2020.

5.1 Training and Test Dataset

Figure 6 represents the mean and standard deviation of the two signals monitored over time. The noticeable drop in the signals (highlighted by a vertical red line in the figure) occurs at the same time as some maintenance work was performed due to an oil leakage problem. It appears that the maintenance interfered with the sensors, and data collection stopped working correctly after the intervention. Therefore, measurements collected after this date were excluded from further analysis.

The remaining measurements were used as shown in Fig. 7. The data collected between December and May (which represent an entire life cycle of the machinery) were used to train the predictive model, while the measurements collected between October and December were used to test the goodness of our model. In this way, 30491 production cycles were used in the training phase, while 16821 production cycles were used with new data on which to test the model.

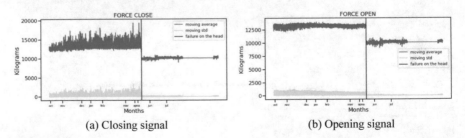

(a) Closing signal   (b) Opening signal

Fig. 6 Mean value and standard deviation of the signals over time. The red vertical line marks a maintenance intervention, after which the sensor appear to not functioning properly. Only data acquired before the intervention were included in the analysis

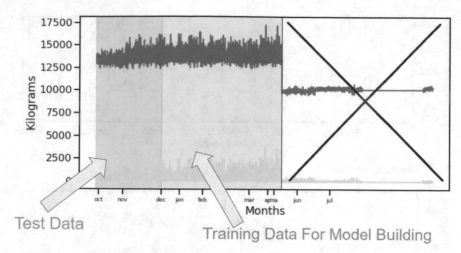

Fig. 7 Definition of the training and test data for the RUL model

5.2 Experimental Settings

Each production cycle consists of 1000 different measurements. The preprocessing step extracts statistical features that summarise the information content of these measurements in a more compact representation. Based on the advice of a domain expert, we divided each production cycle into 12 segments of equal size. For each of these segments, the 14 features listed in the Sect. 5 were then calculated, resulting in 168 variables per cycle. Only features that were not highly correlated with other variables were actually considered in the model building. Feature selection is applied with a 0.5 correlation threshold, that means that features with a Mean Absolute Correlation value greater than 0.5 were discarded. The model was built on the remaining 81 features.

The next steps are defining the size and step of the sliding time window. Given that an average of about 500 production cycles are performed in a working day, the window size was set to 2500 cycles (about 5 working days). In this way, sufficient data are available to compare the distribution of the features between two time windows. The window is moved by 1250 cycles at each step, to ensure that the model is able to perform a sufficient number of comparisons in the training phase. With these settings, the second half of one window will be equal to the first half of the following window.

5.3 Model Building and Test

Following Sect. 5.1, the first 2500 production cycles starting from December 2019 (the beginning of the machine life cycle) were used as the baseline window. Then,

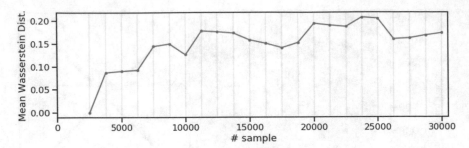

Fig. 8 Distances over time

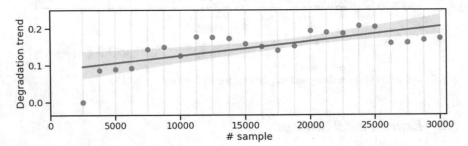

Fig. 9 Model trend

all subsequent windows are compared to this one and the distance between them is calculated.

Figure 8 shows the plot of these distances over time. Each point represents the distance between the preceding 2500 production cycles and the reference window. The first point is equal to 0 because the reference window was compared to itself. From the subsequent comparisons, the distance starts to increase, a sign that the machine is degrading over time. The degradation process is modelled by a linear regression of the calculated distances, as shown in Fig. 9.

The regression model identifies an increasing trend, in accordance with what we expected. The linear fit is obtained during an entire life cycle of the machine. When new measurements are collected, the predictive model will be able to estimate the degradation of the machine, based on those measurements.

The model was tested on the data collected before December 2019 (which were not included in the training phase); specifically, 16821 production cycles were collected between October and December. It is important to note that these measurements represent the end of the life cycle that precedes the one used for training. These measurements end at the beginning of December with a maintenance intervention due to a fault on the machine. For this reason, it is reasonable to presume that the degradation of the machine at that time was quite high.

The model was tested on two different windows, highlighted in red in Fig. 10. The first window contains measurements collected between October and November (about one month before the failure), the second window contains measurements

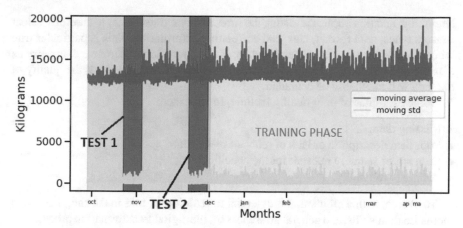

Fig. 10 Test time windows used to evaluate the model are highlighted in red

Table 1 Estimated degradation

	Machine degradation (%)	Remaining cycles
Test 1	77.59	7392
Test 2	100	0

collected at the end of November, close to the replacement. Both windows contain 2500 measurements (same as the reference window).

The results are shown in Table 1. For each test window, we measured the percentage of machine degradation and the estimated RUL, expressed as the number of life cycles that the machine is still capable of performing before maintenance. The results are in line with expectations. For the first time window (Test 1), a machine degradation of almost 78% and a RUL of 7392 cycles remaining was estimated. In reality, the machine broke down after 9429 production cycle. As regards the second time window (Test 2), a 100% machine degradation was estimated. These production cycles immediately preceded the maintenance intervention and were indeed measured at the end of the machine life cycle. Considering the failure that occurred shortly afterwards, both estimates are reasonable and not too distance from observations.

6 Lessons Learned

Beside practical results, SERENA provided some important lessons to be taken into account when building predictive maintenance services in the white goods industry.

Data Quality. Finding the relevant piece of information hidden in large amounts of data turned out to be more difficult than initially thought. One of the main lessons learned is that Data Quality needs to be ingrained in the process from the very beginning, and this implies investing the time, effort and resources to carefully select sensors types, data format, tags, and correlating information. This is particular true when dealing with human-generated data: if operators do not perceive data entry as a useful and impactful activity, rather than a waste of time and energy, the quality of the data will inevitably be degraded.

Some examples of poor quality include, for instance:

- Missing data;
- Poor data description or lack of relevant metadata;
- Data not or scarcely relevant for the specific need;
- Poor data reliability.

To overcome this problem, it is essential to **train operators** in the shop floor, and increased their skills, and general awareness of, the digital transformation process and specifically data-driven decision processes. It is likewise important to design more **ergonomic human machine interfaces**, involving experts in the field to reduce the time and mental effort needed for data entry and lower the possibility of errors and inconsistencies.

Following these two recommendations, it is possible to design better datasets from the beginning, which in turn ensures higher quality and availability of machine-generated data, as well as reduce the possibility of errors, omissions and scarce accuracy in human-generated data.

It is worth stressing that, in order to be able to train machine learning models on large, rich, and representative datasets, data collection needs to be designed in advance, months, even years before the real need may emerge. This can be achieved by following some simple guidelines:

1. **Anticipate the installation of sensors and data gathering**. The best is doing it at the equipment first installation or at its first revamp activity. The amount of data needed to build accurate and reliable machine learning models should not be underestimated. A financial plan should be devised, as investments in equipment (sensors) and data storage will be recouped by cost savings after some years.
2. **Gather more data than needed**. A common practice is to design a data gathering campaign starting from an immediate need. This could lead however to leaving out critical data when future needs emerge, therefore wasting resources in new data gathering campaigns. Thus, a well-thought campaign may allow, with a modest increase in time and effort, to support multiple applications and needs. In an ideal state of infinite capacity, data gathering activities should be to capture the full ontological description of the system under design. Of course, this could not be feasible in real-life situations. A good intermediate strategy could be to populate the machine with as many sensors as possible.
3. Start initiatives to **preserve and improve currently available datasets**, even if not immediately needed. For example, loose spreadsheet files spread in individuals' PCs should be migrated into a common shared database, making sure that the data is cleaned and normalised (e.g., by converting local languages descriptions in data and metadata to English).

Skills. Data Scientists and Process Experts are not (yet) "talking the same language" and it takes significant time and effort from mediators to make them communicate properly. This is also an aspect that needs to be taken into account and carefully planned. Companies definitely need to close these "skills" gaps. Among the various strategies that could be applied, it is possible to train process experts on data science; train data scientists on the subject matter; or develop a new Mediator role, which shares the minimum common ground to act as a facilitator between the two profiles.

7 Conclusions

The work proposed in this chapter represents a SERENA use case in which the platform is exploited to address a maintenance predictive problem in industrial foaming machines. The SERENA architecture based on a lightweight micro-services architecture proved to be effective also in this use case. The specific challenge addressed consisted in the definition of a new methodology able to evaluate the degradation of the machinery, and estimate its remaining life in terms of production cycles that can still be carried out. The results obtained with the first test conducted are promising and indicative of the reliability of the methodology. The proposed approach is a principled and general methodology, based on simple yet elegant theoretical principles, which can be applied also to many other white goods industry scenarios.

Acknowledgements This research has been partially funded by the European project "SERENA—VerSatilE plug-and-playplatform enabling REmote predictive mainteNAnce" (Grant Agreement: 767561).

References

1. S. Wang, J. Wan, D. Zhang, D. Li, C. Zhang, Towards smart factory for industry 4.0: a self-organized multi-agent system with big data based feedback and coordination. Comput. Netw. **101**, 158 (2016)
2. Y. Chen, F. Zhu, J. Lee, Data quality evaluation and improvement for prognostic modeling using visual assessment based data partitioning method. Computers in Industry **64**(3), 214 (2013)
3. J. Dalzochio, R. Kunst, E. Pignaton de Freitas, A. Binotto, S. Sanyal, J. Favilla, J. Barbosa, Machine learning and reasoning for predictive maintenance in industry 4.0: current status and challenges. Comput. Indus. **123**, 103298 (2020). https://doi.org/10.1016/j.compind.2020.103298
4. M. Motahari-Nezhad, S.M. Jafari, Bearing remaining useful life prediction under starved lubricating condition using time domain acoustic emission signal processing. Exp. Syst. Appl. **168** (2021). http://www.sciencedirect.com/science/article/pii/S0957417420310630
5. T. Cerquitelli, N. Nikolakis, P. Bethaz, S. Panicucci, F. Ventura, E. Macii, S. Andolina, A. Marguglio, K. Alexopoulos, P. Petrali et al., Enabling predictive analytics for smart manufacturing through an iiot platform. IFAC-PapersOnLine **53**(3), 179 (2020)

6. S. Proto, F. Ventura, D. Apiletti, T. Cerquitelli, E. Baralis, E. Macii, A. Macii, Premises, a scalable data-driven service to predict alarms in slowly-degrading multi-cycle industrial processes, in *2019 IEEE International Congress on Big Data, BigData Congress 2019, Milan, Italy, July 8-13, 2019*, ed. by E. Bertino, C.K. Chang, P. Chen, E. Damiani, M. Goul, K. Oyama (IEEE, 2019), pp. 139–143
7. J. Lee, H.A. Kao, S. Yang, Service innovation and smart analytics for Industry 4.0 and big data environment. Procedia CIRP **16**, 3 (2014)
8. D. Apiletti, C. Barberis, T. Cerquitelli, A. Macii, E. Macii, M. Poncino, F. Ventura, istep, an integrated self-tuning engine for predictive maintenance in industry 4.0, in *IEEE ISPA/IUCC/BDCloud/SocialCom/SustainCom 2018, Melbourne, Australia, December 11–13, 2018* (2018), pp. 924–931
9. L. Rüschendorf, The Wasserstein distance and approximation theorems. Prob. Theor. Relat. Fields **70**(1), 117 (1985)
10. W.W.W. Consortium, et al., JSON-LD 1.0: a JSON-based serialization for linked data (2014)
11. MIMOSA 2020. www.mimosa.org

Predictive Analytics in the Production of Elevators

Valeria Boldosova, Jani Hietala, Jari Pakkala, Riku Salokangas,
Petri Kaarmila, and Eetu Puranen

Abstract With the emerging role of digitalization in the industrial sector, more
and more companies attempt to increase asset availability, improve product quality
and reduce maintenance costs. Manufacturing companies are faced with the need
to transform traditional services into remote factory monitoring solutions using big
data and advanced analytics. Kone is a global leader in the elevator and escalator
production industry, which is continuously looking for new ways of improving production efficiency and reducing machine downtime in order to run unmanned 24/7
production. However, the process of collecting data from equipment and utilizing
it for predictive analytics can be challenging and time consuming. Therefore, during Serena project Kone cooperated with VTT and Prima Power, which provided
necessary capabilities and competencies in the areas of data collection, analysis and

V. Boldosova (✉) · J. Pakkala
Prima Power, Nuppiväylä 7, 60100 Seinäjoki, Finland
e-mail: valeria.boldosova@primapower.com

J. Pakkala
e-mail: jari.pakkala@primapower.com

V. Boldosova
University of Vaasa, Wolffintie 34, 65200 Vaasa, Finland
e-mail: valeria.boldosova@student.uwasa.fi

J. Hietala · R. Salokangas · P. Kaarmila
VTT Technical Research Centre of Finland, Kemistintie 3, 02150 Espoo, Finland
e-mail: jani.hietala@vtt.fi

R. Salokangas
e-mail: riku.salokangas@vtt.fi

P. Kaarmila
e-mail: petri.kaarmila@vtt.fi

E. Puranen
KONE, Hissikatu 3, 05830 Hyvinkää, Finland
e-mail: eetu.puranen@kone.com

© Springer Nature Singapore Pte Ltd. 2021
T. Cerquitelli et al. (eds.), *Predictive Maintenance in Smart Factories*,
Information Fusion and Data Science,
https://doi.org/10.1007/978-981-16-2940-2_8

utilization for developing and testing predictive maintenance solutions in the elevator manufacturing industry. As a result of this collaboration, VTT integrated sensors into Prima Power production line used at Kone and developed algorithms for measuring the remaining useful life of conveyor bearings. As a machine tool builder, Prima Power contributed to the project with a cloud environment for remote collection of vibration measurement data and Serena Customer Web analytics for condition-based maintenance.

1 Challenges and Needs in Elevator Production Industry

Trying to survive in times of competitive business environment, high uncertainty and low growth, manufacturing companies are looking for new opportunities to improve fleet reliability and reduce maintenance costs. The proliferation of connected machines, sensors, and advanced analytics in the age of digitalization creates multiple benefits for companies that know how to use them. In the elevator production industry, Kone is a leading manufacturing company with more than 50,000 employees and operating globally across 60 countries, which is eager to take advantage of digitalization and increase equipment availability through predictive maintenance.

Manufacturing of elevator cabins involves sheet metal processing through punching and bending. The machinery that Kone utilizes on the factory floor is provided by Prima Power, a large international provider of sheet metal processing machines and software. For the production of elevators Kone uses automated PSBB (Punching-Shearing-Buffering-Bending) manufacturing system to automatically process blank metal sheets into ready-bent, high-quality components. The summary of the production stages at Kone using Prima Power machinery is illustrated in Fig. 1.

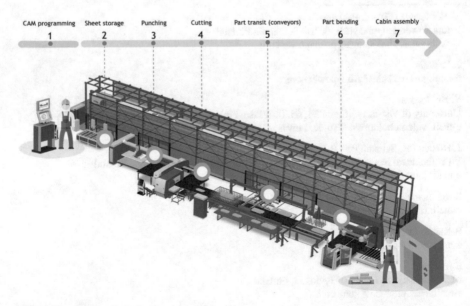

Fig. 1 Production process of elevator cabins at Kone

Due to the large-scale production of elevator cabins and demanding deadlines, Kone often has to run the production line unmanned during the night shifts. Therefore, the reliability of the manufacturing line during the evening and night shifts is crucial to the production effectiveness at Kone.

Based on the detailed Failure Mode and Effect Analysis (FMECA) conducted during the project, conveyor bearing and punching tooling have been identified as critical components at Kone that affect performance of the production line and require additional attention. The unexpected wear or failure of these components can reduce part quality and lead to production interruption until the problem is resolved. However, these challenges can be solved by remotely monitoring the condition of conveyor bearings, predicting wear and scheduling preventive maintenance before the actual failure. Since data collection, analysis and utilization can be challenging tasks, Kone collaborated together with Prima Power and VTT Technical Research Centre during Serena project to combine expertise, capabilities and know-how.

During the project, Kone acted as a pilot in elevator manufacturing industry providing the real-time environment for testing predictive maintenance prototypes at an operating production facility in Finland. The collaboration between Kone, VTT Technical Research Centre and Prima Power involved data collection through installed bearing vibration sensors, punching tool acoustic emission sensors, and microphones in the production line (Fig. 2). Machine vibration and excessive noise are often related to the worn parts, and installed sensors help to detect excess vibration before it causes unplanned downtime. Data collected through sensors has been stored in Prima Power cloud and then further utilized to develop predictive maintenance algorithms and models in Serena Customer Web platform to accurately predict failures.

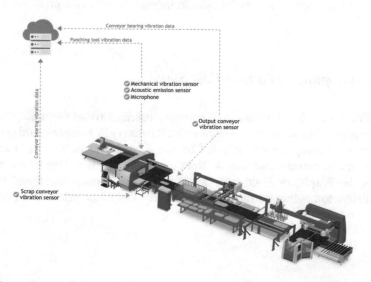

Fig. 2 Data collection from the PSBB production line

The overarching project goal for Kone, VTT and Prima Power was to gain an in-depth understanding of how to use big data to predict component life expectancy and forecast when conveyor bearings have to be replaced or when punching tools will need grinding in order to maintain high part quality. As a result, the planning and scheduling of preemptive maintenance activities as well as spare part ordering can be done in advance without interrupting the production process. In addition to improving the technical availability of the production line, maintenance costs can be reduced due to shortened waiting and repair time. Instead of a technician physically checking the condition of the production line on-site, it can be now monitored remotely in real-time.

2 Sensors and Data Acquisition Devices for Remote Condition Monitoring

During Serena project, VTT utilized low-cost components to develop the condition monitoring system with sensors and edge processing. One of the benefits of the low-cost solution for condition monitoring is that it can be easily implemented on a large-scale. Furthermore, maintaining a low-cost edge device requires fewer expenses since the part replacement is inexpensive. On the other hand, low-cost devices are often prone to failure over time and demand extra maintenance in comparison with the more expensive counterparts. Additionally, low-cost parts can lead to unforeseen development costs, for example, if the hardware components require additional tuning or device interfaces are not well developed. Also, lower cost devices might not be able to handle high sampling rates. To avoid the above mentioned pitfalls, during Serena project VTT decided to explore both popular and also less known system components to ensure a successful system implementation in a production line at Kone.

2.1 Development of a Low-Cost Solution

The collection of vibration data from conveyor bearings (from the rolling element) and edge processing were implemented with a Raspberry Pi 3 single-board computer. This type of a computer was selected for the Serena project because it can take vibration measurements and also do light processing of data. One of the benefits of using the Raspberry Pi edge device is that it has a great wired and wireless connectivity, computing performance, and a relatively low price. However, even though the Raspberry Pi can connect to various sensors via its General-Purpose Input Output (GPIO) pins, it is only capable of collecting digital signals. Therefore, an external Analog to Digital Converter (ADC) is required for reading data from analog sensors [1].

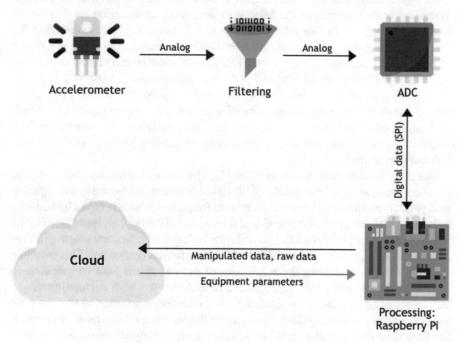

Fig. 3 Measurement signal travel path within the low-cost measurement system built for vibration data collection from a conveyor bearing

As a result of Serena project, the developed measurement system consists of an accelerometer, a low-pass filter, an ADC, and the Raspberry Pi (Fig. 3). The accelerometer produces an analog vibration signal, which is then processed with a low-pass filter to avoid the aliasing effect. Aliasing refers to the signal distortion in signal processing due to a sampling frequency that is too low for the signal bandwidth. When sampling an analog signal, the sampling frequency should be at least twice the frequency of the signal to be sampled according to Nyquist's theorem. Next, the analog signal is converted to a digital signal in the ADC and the data is communicated to the Raspberry Pi via Serial Peripheral Interface (SPI). Finally, the Raspberry Pi processes and transfers both processed and raw data to the cloud. In addition to this, the Raspberry Pi receives equipment parameters back from the cloud [2].

2.2 Edge Processing

During the project, the edge analytics approach to data collection and analysis was selected because it ensures that data is processed and analyzed as close as possible to the object from which the information is collected. One of the benefits of edge processing is that it reduces the latency in data analysis enabling near real-time decision-making. For further information about edge analytics, please see Chap. 3.

In the bearing diagnostics, there are four major fault modes: outer and inner race, the rolling element, and the cage or train [3]. Each of these bearing faults has its own distinctive nominal pulse interval and one such example is demonstrated in Fig. 4. These pulse intervals are determined by the rotating speed and the geometry of the bearing. Once the expected frequencies of the fault pulse intervals are calculated, the analysis can focus only on these fault frequencies. The amplitudes of the fault frequencies can then be monitored to determine the bearing health condition at each different part of the bearing [4].

One of the best methods to observe these fault frequencies is the envelope analysis. In the envelope analysis the unfiltered vibration acceleration time domain signal is band pass filtered around the bearing's natural frequency, which is typically between 500 and 3000 Hz, depending on the size of the bearing [3]. After band pass filtering the time domain signal is rectified and demodulated using the Hilbert transform [5]. This produces an envelope signal for which the FFT transform is carried out. This envelope spectrum is used to monitor the amplitudes of pre-calculated fault frequencies. In this particular use case, the edge processing includes the both calculation of the fault frequencies and data manipulation with the envelope analysis so that the fault frequency amplitudes can be found from under the superimposed random vibrations. The fault frequency calculation is included to the edge analytics because these fault frequencies can be changed when the rotation speed of the bearing changes. During the envelope analysis, data was processed with the Raspberry Pi as follows. First, a band-pass filter was applied to the data to attenuate interfering signals caused by

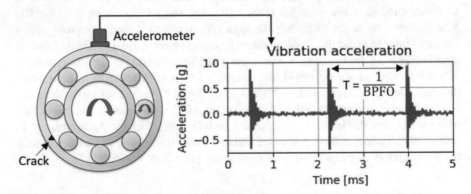

Fig. 4 A crack in the outer race of a bearing is observed as repeated ringing pulses. The interval between the pulses corresponds to the ball pass frequency on the outer race (BPFO) [4]

misalignment and imbalance and to eliminate the random noise outside the passband. In this special use case the goal was to identify the amplitudes of bearing fault frequencies and not to focus on misalignment and imbalance this time. After filtering, the ringing pulses were rectified with Hilbert transform [5] and calculated into a spectrum, which then revealed the pulses at fault frequencies. Finally, the energies of the identified nominal frequency peaks were computed [4].

2.3 Use Case Results

Within the Serena project, VTT conducted two experiments by using the developed low-cost system to test the measurement accuracy and viability of the solution. First, the system was tested with a hand calibrator oscillating at set frequency of 160 Hz. Figures 5 and 6 show promising results as the low-cost system evenly responded to the generated vibration signal and was able to correctly detect the frequency of the hand calibrator in the spectrum [2].

Second, the system was compared to a higher cost sensor and data acquisition device, which offer better accuracy and signal-to-noise ratio. In this experiment, the data was analyzed using VTT O&M Analytics creating a spectrum of envelope signal from which the theoretical fault frequency magnitudes of the bearing can be

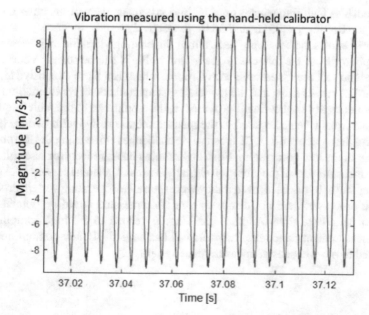

Fig. 5 Vibration of the hand calibrator vibrating at 160 Hz in time domain as measured by the low-cost system [2]

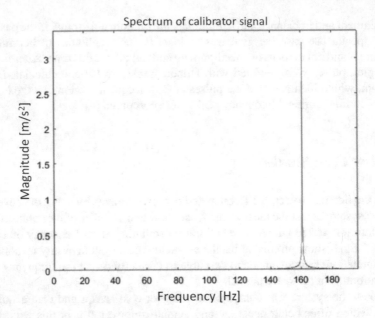

Fig. 6 Spectrum of the vibration created with the hand calibrator vibrating at 160 Hz as measured by the low-cost system [2]

compared. The fault frequencies used in this experiment were of the outer and inner rings of the bearings, at 25 Hz and 38 Hz respectively.

Figure 7 shows the envelope spectrum of the bearing vibration as measured with the Raspberry Pi of the low-cost system, and in Fig. 8 can be seen the corresponding envelope spectrum measured with Cronos, which is a more expensive technical solution. The results show that the low-cost system could detect both outer and inner fault frequencies of the bearing. Furthermore, the amplitudes of the fault frequencies are almost equal when compared to the measurements created with Cronos [2].

During the experiments, VTT faced some challenges while using the Raspberry Pi 3 as a part of the low-cost system. First, any running background processes interfered with the measurement of the vibration signal. As a result, all unnecessary background processes had to be disabled, even if they were the default operating system processes.

If the processes were not disabled, a reliable measure could be obtained only for short periods of time. Another challenge is related to the high sampling rate of the measurements since the Raspberry Pi has a relatively low computing power compared to more expensive measurement systems [2].

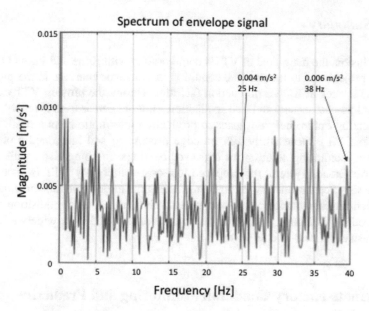

Fig. 7 Envelope spectrum of a conveyor bearing vibration signal. Created with the low-cost system and analysed by VTT O&M Analytics. Low acceleration amplitude at outer (25 Hz) and inner (38 Hz) race fault frequencies indicate that the bearing is in good condition [2]

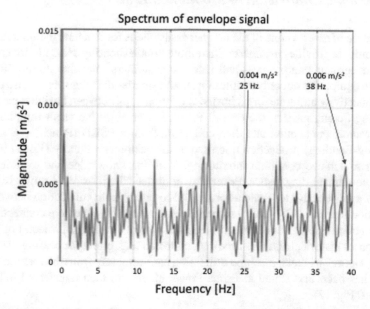

Fig. 8 Envelope spectrum of a conveyor bearing vibration signal. Created with Cronos and analysed by VTT O&M Analytics. Low acceleration amplitude at outer (25 Hz) and inner (38 Hz) race fault frequencies indicate that the bearing is in good condition [2]

2.4 Summary

In conclusion, the main goal of VTT's collaboration with Kone and Prima Power in Serena project was to monitor the condition of conveyor bearings in the punching machine located at KONE production facilities. During the project, VTT explored whether low-cost sensors and data acquisition solutions can be used to reliably determine bearing failure before it leads to production interruption. As a result of collaboration, VTT successfully utilized edge processing and integrated a proactive condition monitoring solution for conveyor bearings. In comparison with a more expensive Cronos system, the proposed low-cost solution by VTT is expected to provide significant monetary savings considering the large-scale implementation by manufacturing companies.[1] Low-cost sensor solutions and data acquisition devices developed by VTT have promising results and are expected to be widely utilized in the industry in a cost-effective way.

3 Remote Factory Condition Monitoring and Predictive Analytics

3.1 Remote Connection to Production Line

Digitization of assets is one of the first necessary steps for machine manufacturers to take in order to provide customers with remote troubleshooting and data-driven products, reduce maintenance costs and increase customers' machine uptime. Remote access to equipment not only enables faster and smarter decisions, but it also creates interconnectivity and transforms business relationships between business partners.

During Serena project, the overall Prima Power objective was to implement a secure remote connection and data collection from a PSBB (Punching-Shearing-Buffering-Bending) production line used at the customer's factory (Kone) (Fig. 1.) in order to achieve condition monitoring. With the knowledge and experience in remote equipment diagnostics, Prima Power utilized Tosibox VPN router [6] (supplied by a third party) to create a secure end-to-end remote connection between the production line and PC. Figure 9 summarizes the established remote connectivity to the production line at Kone through a Tosibox device during the project. For practical purposes historical and real-time data collected (e.g. conveyor bearing, punching tooling, alarms, triggers, machine utilization etc.) remotely from the manufacturing system has been first stored in Prima Power cloud and then transferred to Serena cloud (cf. Fig. 17).

---

[1] The exact savings from using the low-cost solution in comparison with the more expensive devices are difficult to accurately estimate due to large variation in equipment pricing levels and different levels of system implementation into the factory environment.

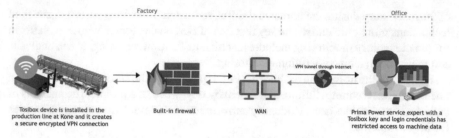

Fig. 9 Remote connection to a production line

From the perspective of data privacy and security, Tosibox VPN router is certified according to ISO 27001:2013 and IEC 15408 standards. Tosibox remote connection device has a built-in firewall, and the data processing procedure complies with EU General Data Protection Regulation (GDPR) 2016/679 [7]. Access to the data collected from the production line at Kone and stored in Prima Power cloud is granted only to authorized persons with valid login credentials and a Tosibox key, a cryptoprocessing device. Tosibox key is a physical USB security key that provides user access to the collected data through safe encrypted VPN connection. The user permissions to access data can be easily granted or revoked by the organization that offers remote connection and condition monitoring service (i.e. Prima Power).

3.2 Versatile Cloud-Based Platform for Remote Diagnostics

As a manufacturer of sheet metal processing machinery, Prima Power competitiveness on the market depends whether it can expand software portfolio and create new predictive maintenance and condition monitoring analytics that provide added value and satisfy customer needs. During Serena project, in addition to developing a solution for remote connectivity and data collection from the production line at Kone, Prima Power also built SaaS (software as a service) application on Microsoft Azure for analytics based on the captured data. Serena Customer Web is an instance from the Serena cloud and a cloud-based platform that contains a number of dashboards with predictive analytics for manufacturing companies to remotely monitor and optimize performance of Prima Power machines. Within Serena project, the developed solution has been piloted and tested together with Kone with respect to usability and user experience.

Serena Customer Web dashboards provide manufacturing companies (e.g. Kone) with machine historical data, real-time diagnostics, AI condition-based maintenance, service scheduling, AR-based assistance for operators and recommendations how to increase machine performance. Serena Customer Web helps to predict downtime and notifies users in advance when to schedule service maintenance, order spare parts and replace components. As a result, Serena Customer Web helps to increase reliability of the machinery during the unmanned shifts, reduce maintenance costs and use field technicians more efficiently. The key features of Serena Customer Web as a platform for remote factory monitoring include: machine utilization reporting, alarm analysis, and real-time machine condition monitoring.

First, machine utilization dashboard (Fig. 10) provides the summary of production line utilization (running, failure, idle times) by day, week or a month. The analysis of historical data enables users to identify performance trends and uncover opportunities for improvements.

Then, alarm analysis dashboard (Fig. 11) gives the overview of frequent machine alarms showing which alarms are causing most of production downtime during the day, evening and night shifts. The purpose of this dashboard is to detect patterns and abnormalities, inform the manufacturing company and advise how to modify future production operations.

Finally, real-time machine condition dashboard (Fig. 12) demonstrates machine operations, performance and alarms in real-time to help diagnose and resolve bottlenecks during the day, evening or night shifts. This dashboard provides an overview of manual operations and alarms prior to the interrupted production process so it helps users to identify what triggered an alarm and caused unexpected failure. When the alarm is triggered, Serena Customer Web video server records short 2 min. video (through cameras installed in the machine) of what was happening prior to machine interruption. Watching these videos supports users in identifying the root cause of downtime.

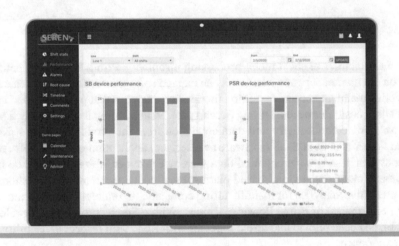

Fig. 10 Serena Customer Web dashboard: machine utilization

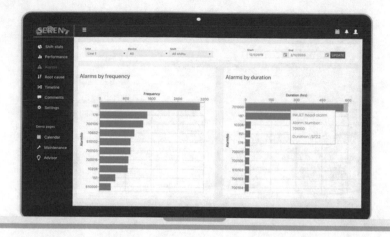

Fig. 11 Serena Customer Web dashboard: alarm analysis

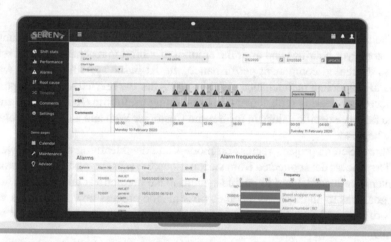

Fig. 12 Serena Customer Web dashboard: real-time machine condition

3.3 *Predictive AI Condition-Based Maintenance and Service Scheduling*

Serena Customer Web does not only monitor the actual condition of the production line but it also predicts which maintenance should be done and when in order to prevent unexpected machine downtime. In particular, during the project, Serena Customer Web was remotely collecting the conveyor bearing vibration data through the installed sensors (by VTT) in PSBB line at Kone factory and it was programmed to proactively schedule a service event in the calendar (Fig. 13) if it detected the wear of the components.

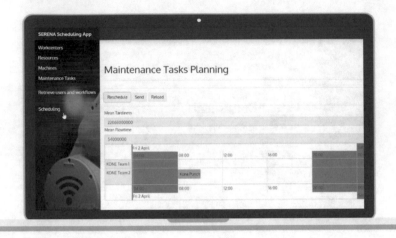

Fig. 13 Serena Customer Web dashboard: service scheduler

The platform was developed in such way that it is continuously analyzing data from conveyor bearings and after the amount of vibration reaches a critical set limit due to decreased component performance, the AI-based system reacts by warning Kone to schedule a maintenance visit shortly. After the service technician performs the necessary bearing replacement, the remaining useful life (RUL) values are reset to the factory-specific values in the Serena Customer Web and a new data collection cycle begins.

During Serena project, regression models and deep learning principles were utilized to process data collected from various sources (sensors, past failure history and maintenance records) and to enable system make predictions about future failures. The goal was to automate as much as possible the process how AI collects various types of data and learns from it over time to determine the optimal time for service activity.

In contrast to the old-fashioned manually scheduled periodical maintenance, AI condition-based maintenance tool in Serena Customer Web is more efficient (see forthcoming 'Impact' section) since the maintenance is performed only on as-needed basis and only when the AI detects the decrease in the performance of the equipment. The condition monitoring and scheduling are performed while the production line is running without disrupting the daily operations in Kone, thus reducing average time the technician spends on maintenance.

3.4   AR-Based Assistance for Operators During Maintenance

In addition to the condition-based maintenance tool and proactive service scheduling, Serena Customer Web also helps operators to do maintenance work on the factory floor faster and more efficiently by providing them with the practical situation-specific assistance (Fig. 14).

As a result of Serena project, the interactive animated guidance showing operator how to replace tooling in Prima Power machine has been developed in cooperation with SynArea (Fig. 15), while the tool grinding step-by-step video instructions have been designed together with Oculavis (Fig. 16).[2]

Both types of manuals available in Serena Customer Web have been piloted and tested together with Kone during the project. The tooling replacement guidance (Fig. 15) provides user with a 3D model of the punching machine and assists operator during the process through animated sequences for each step (e.g. open panel, lift tool, loosen screws). The tool grinding guidance (Fig. 16) contains brief video instructions for each step of the maintenance procedure that support operator from start to finish. Regardless of the operator's knowledge in maintaining the production line, Serena Customer Web manuals are designed to be easy to understand and can be used by both novice and expert field technicians. Digitalized guidance replaces paper-based manuals and takes maintenance to the next level by assisting operators through visual and voice commands, hence improving safety on the factory floor and reducing human errors. In the future (outside the Serena project scope), the AR-based assistance can be even further enhanced with the help of hands-free wearable

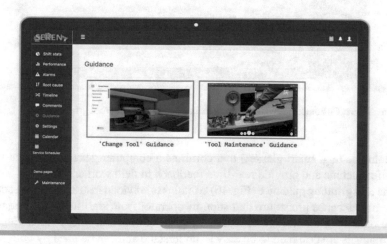

Fig. 14   Serena Customer Web dashboard: operator's guidance

---

[2] SynArea AR guidance and Oculavis video instructions can be accessed only by project partners with the valid login credentials to Serena Customer Web. However, to enhance readers' understanding of these functionalities, their screenshots are provided in Figs. 14, 15 and 16.

Fig. 15 Serena Customer Web dashboard: guiding operator during tool replacement

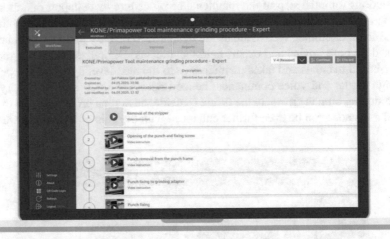

Fig. 16 Serena Customer Web dashboard: guiding operator during tool grinding

technologies (e.g. smart glasses) that combine a computer-generated image with a real-life machine and provide real-time feedback to field workers.

The tool grinding guidance (Fig. 16) contains brief video instructions for each step of the maintenance procedure that support operator from start to finish. Regardless of the operator's knowledge in maintaining the production line, Serena Customer Web manuals are designed to be easy to understand and can be used by both novice and expert field technicians. Digitalized guidance replaces paper-based manuals and takes maintenance to the next level by assisting operators also through the hands-free wearable technologies (e.g. smart glasses) combining a computer-generated image with a real-life machine for improved safety on the factory floor and reduced human errors.

4 Integrated Solution and Technical Architecture

The overall technical infrastructure for data collection and data processing, which was developed and utilized by Kone, VTT and Prima Power during Serena project is illustrated in Fig. 17. Real-time data was collected from the production line at Kone using the Raspberry Pi and a Tosibox device to enable remote factory condition monitoring. Data was first stored in the Prima Power cloud and then it was moved to the Serena cloud as demonstrated in the figure below. Serena Customer Web platform (repeatedly mentioned in this book chapter) is a software solution for remote factory condition monitoring, which is based on Serena cloud and it provides an overview of customer's (i.e. Kone) data processed with big data analytics and presented through various visualization tools (Figs. 10, 11, 12, 13, 14, 15 and 16).

Serena Customer Web platform was designed in such way that customers can get access only to their own production data, however because of the software scalability the same cloud environment can be replicated and extended for other customers. As a result, the supplier of sheet metal processing machinery (i.e. Prima Power) can provide remote condition monitoring and predictive maintenance solutions not only for Kone, but to the entire installed customer base around the world.

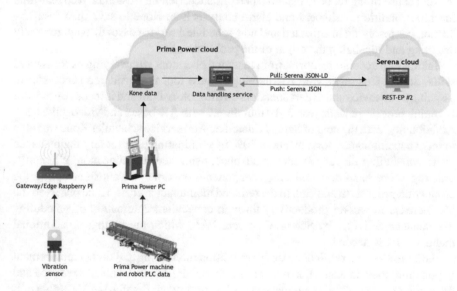

Fig. 17 Technical infrastructure

5 Impact

The benefits of close collaboration between Kone, VTT and Prima Power are twofold, because this kind of cooperation gave new insights to both industrial companies as well as the research partners involved in the project (including SynArea, LMS, Oculavis). Serena project results transformed business relationships between partners and helped to gain competitive advantage over competitors through a unique opportunity to develop and integrate predictive maintenance solutions in elevator production industry. For example, based on the KPIs measured in the beginning of Serena project and upon project ending it can be seen that Kone gained significant cost savings in the elevator production process in comparison with other players on the market (cost advantage). On the other hand, Prima Power followed the differentiation path, increased customer loyalty and gained a unique position in sheet metal processing industry by expanding product portfolio with predictive factory monitoring solutions. More detailed information on successful Serena project outcomes for Kone, VTT and Prima Power are provided in the rest of this section.[3]

As a result of Serena project, Kone has been able to observe an approximately 5% increase in technical availability through reduced failure rate and time spent during the repair. While, it can be challenging and time consuming to physically identify the root cause of the reduced machine performance, Serena Customer Web real-time machine condition dashboard and alarm analysis help Kone to save time (est. 15–30 min. per resolving unexpected and non-scheduled service case) through remotely detecting and localizing the origin of the problem online.

Taking into account feedback from machine operators with varying expertise and skills, Serena Customer Web AR-based guidance tool is considered successful in assisting both novice and expert operators on the factory floor and it speeds up the tool changing process in Kone (est. 5–10 min. faster with AR-based and video guidance). Additionally, with the help of Serena Customer Web service scheduler Kone was able to reduce maintenance costs by about 10% by eliminating unnecessary maintenance visits, scheduling preventive care in advance, using service labor more efficiently, ordering spare parts and replacing components on time without interrupting the production process. In addition to the reduced maintenance costs, Kone observed the 5% increase in worker productivity through convenient automation of scheduling maintenance activities by Serena Customer Web, which means that no additional manual work is needed.

Furthermore, as a result of proactive condition monitoring and timely replacement of punching tools in Kone, the number of material defects and damages reduced and the overall quality of the sheet metal parts has improved. Finally, the VTT data collection devices and Serena Customer Web developed by Prima Power during Serena project increased the overall job satisfaction (based on the received oral feedback) of Kone operators' and managers' with the performance of the production line by

---

[3] The exact performance indicators cannot be disclosed in this chapter due to strict confidentiality policies in participating companies. Nevertheless, to ensure that readers get an overall understanding of the successful project outcomes the key metrics are discussed in relative terms in this section.

simplifying maintenance procedures and reducing costs. Summary of measurable KPI values collected from Kone production facilities before and after implementing Serena testbed is demonstrated in Fig. 18.

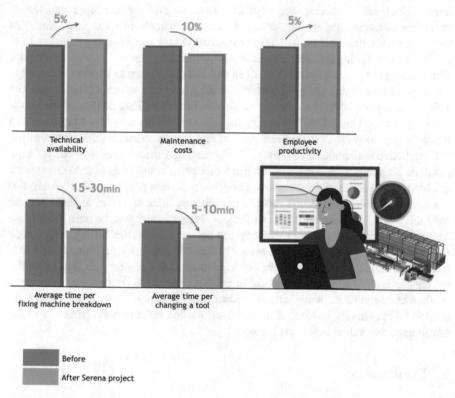

Fig. 18 Impact of Serena project on Kone elevator production

Despite the already visible positive impact of predictive maintenance solutions on elevator manufacturing industry, it should be acknowledged that the estimates provided in Fig. 18 were collected during a relatively short period of time and therefore do not 100% accurately reflect the full picture. It is expected that the more realistic numbers and long-term benefits of the project will be seen only after the project completion and after using remote condition factory monitoring solutions for 1–3 years in real factory environment.

The quality of the work done by VTT in Serena project has been demonstrated through numerous high-quality scientific publications. The low-cost solution that VTT implemented during the project has been proved successful in detecting bearing fault frequencies in the same way as a more expensive measurement solution (e.g. Cronos) (with a few small exceptions in the performance quality). From the scientific research perspective, this kind of experimental work hasn't been addressed before in the literature and therefore VTT extends existing academic knowledge by demonstrating the importance of the low-cost solution that provides significant potential savings to manufacturing companies in monitoring the condition of bearings.

In addition to the scientific value of project results, the joint research provided good piloting environment for testing and further commercializing Serena solutions. Serena Customer Web can save days of downtime and prevent large unforeseen costs from unscheduled maintenance for manufacturing companies (e.g. Kone) as well as provide practical monetization implications for machine tool builders (e.g. Prima Power). Taking into account successful piloting of Serena Customer Web with Kone during the project, Prima Power can use this experience as a unique advantage and expand its existing software portfolio with a new data-driven software product Tulus® Analytics beyond the scope of Serena program. This SaaS analytics software can strengthen Prima Power relationship with customers in a long term and it can help to increase customer satisfaction and loyalty. New condition monitoring and predictive maintenance solution will differentiate Prima Power software offering from the competitors and it will strengthen company brand and value to customers. Enabled by the Industrial Internet of Things and Serena project, Tulus® Analytics is a unique software that Prima Power can commercialize and offer to customers all over the world. As a result of Serena project, Prima Power will be able to increase its revenue stream by utilizing a subscription business model and billing customers a recurring fee for a monthly/yearly access to Tulus® Analytics dashboards and analytics. These insights that Prima Power gained from the project can also be useful to other industrial companies seeking to get business value from digitalization and predictive analytics in manufacturing sector. In conclusion, Serena project collaboration has positively influenced all involved parties and provided partner-specific advantages and valuable lessons learned.

6 Conclusion

The teamwork between Kone, VTT and Prima Power in Serena project has been an important milestone in capturing value of big data through the development of next generation predictive maintenance services in the elevator manufacturing industry. Serena project provided a unique environment for industrial companies and research organizations to bring together capabilities, share digitalization vision from different partners, and tackle the revolutionizing of maintenance, which companies couldn't handle separately by themselves. The close collaboration between Kone, VTT, Prima Power and other Serena project partners resulted in effective cross learning and successful know-how sharing.

In addition to the practical industrial implications, Serena project findings provide also valuable insights for the future research initiatives in the area of remote monitoring and predictive analytics beyond the scope of the program. In particular, with the increasing importance of predicting and preventing problems before they occur, there is a constant need to collect a larger variety of data (apart from bearings and tooling) from sheet metal processing machinery to enable uninterrupted, automated and unmanned production during the night shifts. Furthermore, manually created AI predictive models should be further improved to continuously self-adapt to changing conditions, learn and refine themselves over time in order to require minimum human involvement. In order to build the factory of the future, there is a need for self-learning smart machines on the factory floor that can monitor and evaluate their own performance, predict and prevent downtime by automatically ordering spare parts when necessary. As a result, further automating manual and error-prone activities is necessary in order to ensure maximized efficiency in elevator manufacturing industry. At the moment, one of the barriers to successful integration of predictive maintenance in the industrial sector is the high cost of data processing and tailoring AI-based analytics to a particular business context. As a result, future research activities should focus on seeking cost-effective hardware and software solutions for acquiring and analyzing vast volumes of data into actionable insights.

Acknowledgements This research has been partially funded by the European project "SERENA – VerSatilE plug-and-play platform enabling REmote predictive mainteNAnce" (Grant Agreement: 767561).

References

1. Raspberry Pi Foundation, Raspberry Pi 3 model B (2021), https://www.raspberrypi.org/products/raspberry-pi-3-model-b/. Accessed 1 July 2021
2. M. Larrañaga, R. Salokangas, P. Kaarmila, O. Saarela, Low-cost solutions for maintenance with a Raspberry Pi, in *Proceedings of the 30th European Safety and Reliability Conference and the 15th Probabilistic Safety Assessment and Management Conference*, ed. by P. Baraldi, F. Di Maio, E. Zio. 30th European Safety and Reliability Conference ESREL 2020, The 15th Probabilistic Safety Assessment and Management Conference, PSAM 15, Venice, 1–5 November 2020, (Research Publishing, Singapore, 2020), p. 3400, https://www.rpsonline.com.sg/proceedings/esrel2020/html/3780.xml. Accessed 1 July 2021
3. J. Halme, P. Andersson, Rolling contact fatigue and wear fundamentals for rolling bearing diagnostics - state of the art. Proc. Inst. Mech. Eng. Part J J. Eng. Tribol. **224**(4), 377–393 (2010). https://doi.org/10.1243/13506501JET656
4. R. Salokangas, M. Larrañaga, P. Kaarmila, O. Saarela, E. Jantunen, MIMOSA for condition-based maintenance. Int. J. Cond. Monit. Diagn. Eng. Man. **24**(2) (2021). https://apscience.org/comadem/index.php/comadem/article/view/268. Accessed 1 July 2021
5. N.E. Huang, Z. Shen, S.R. Long, M.C. Wu, H.H. Shih, Q. Zheng, N.-C. Yen, C.C. Tung, H.H. Liu, The empirical mode decomposition and the Hilbert spectrum for nonlinear and non-stationary time series analysis. Proc. Roy. Soc. Lond. A Math. Phys. Eng. Sci. **454**(1971), 903–995 (1998). https://doi.org/10.1098/rspa.1998.0193
6. Tosibox, Tosibox - from remote access to secure networking (2021), https://www.tosibox.com/remote-access/. Accessed 1 July 2021
7. Proton Technologies AG, Complete guide to GDPR compliance (2021), https://gdpr.eu/. Accessed 1 July 2021

Predictive Maintenance in the Production of Steel Bars: A Data-Driven Approach

Paolo Bethaz, Xanthi Bampoula, Tania Cerquitelli, Nikolaos Nikolakis, Kosmas Alexopoulos, Enrico Macii, and Peter van Wilgen

Abstract The ever increasing demand for shorter production times and reduced production costs require manufacturing firms to bring down their production costs while preserving a smooth and flexible production process. To this aim, manufacturers could exploit data-driven techniques to monitor and assess equipmen's operational state and anticipate some future failure. Sensor data acquisition, analysis, and correlation can create the equipment's digital footprint and create awareness on it through the entire life cycle allowing the shift from time-based preventive maintenance to predictive maintenance, reducing both maintenance and production costs. In this work, a novel data analytics workflow is proposed combining the evaluation of an asset's degradation over time with a self-assessment loop. The proposed workflow can support real-time analytics at edge devices, thus, addressing the needs of modern cyber-physical production systems for decision-making support at the edge with short response times. A prototype implementation has been evaluated in use cases related to the steel industry.

P. Bethaz · T. Cerquitelli
Department of Control and Computer Engineering, Politecnico di Torino, Turin, Italy
e-mail: paolo.bethaz@polito.it

T. Cerquitelli
e-mail: Tania.Cerquitelli@polito.it

X. Bampoula · N. Nikolakis · K. Alexopoulos (✉)
Laboratory for Manufacturing Systems and Automation, University of Patras, Patras, Greece
e-mail: alexokos@lms.mech.upatras.gr

X. Bampoula
e-mail: baboula@lms.mech.upatras.gr

N. Nikolakis
e-mail: nikolakis@lms.mech.upatras.gr

E. Macii
Interuniversity Department of Regional and Urban Studies and Planning, Politecnico di Torino, Turin, Italy
e-mail: enrico.macii@polito.it

P. van Wilgen
VDL Weweler bv, Apeldoorn, The Netherlands
e-mail: p.van.wilgen@vdlweweler.nl

T. Cerquitelli et al. (eds.), *Predictive Maintenance in Smart Factories*,
Information Fusion and Data Science,
https://doi.org/10.1007/978-981-16-2940-2_9

1 Introduction

Predictive maintenance policies have been used throughout the years, mostly based on human knowledge and intuition as a result of experience. As technology advances, intuition is seeked to be enhanced by computer techniques [1]. The large volume of data generated on a shop floor allow for the use of artificial intelligence techniques that can analyse and create insight over production processes, and as a result complement or support the human knowledge. Nevertheless, selecting the appropriate analysis methods as well as the availability of proper datasets remain a challenge.

VDL Weweler designs, develops, and produces trailing arms, among others, to manufacture trailers, trucks, buses, and cars. The production line of VDL Weweler is fully automated, including both machinery and robots. Maintenance activities, however, are in their great majority either preventive or corrective. Knowing the equipment's working condition and causes of the production interruptions could help identify the maintenance's root cause and restore the system to an operating state. To this aim, proper data-driven predictive maintenance techniques and scheduling for the replacement of segments on the rolling milling machine are discussed and tested. As a result, adequate maintenance planning facilitates further cost reduction and better production management.

We propose and discuss several approaches regarding features extraction and data labelling, considering different subsets of features extracted from the collected data, and labeling the historical set using two different strategies. All combinations between subset of extracted features and data labelling strategy have been tested and evaluated on real data in order to estimate which one is the most performing.

The chapter is organized as follows. Section 2 describes the previous state-of-the-art works present in literature and related to the proposed scenario. Section 3 offers an overview of the Steel production bar production industry, while Sect. 4 details the real-life setting under analysis, focusing on its manufacturing process and the resulting maintenance needs. Then, Sect. 5 describes all the methodology and architecture implemented to provide a data analytics service to the proposed use case, including the obtained experimental results. Finally, Sect. 6 draws conclusions, offering a general summary of what is proposed in this paper.

2 Literature Review

With the introduction of Industry 4.0, smart environments have become very popular, promoting the frequent use of the Cyber-Physical System (CPS), which promotes full integration of manufacturing IT and control systems with physical objects embedded with software and sensors. In this new type of industry, the increased communication between production components leads to a large amount of data. In addition, the integration of Cyber-Physical Systems are encouraging modern industries to transform massive data into valuable knowledge, extracting knowledge about production systems and yielding the support to optimal decision-making [2], helping managers to improve the production processes. The new challenge of modern industries is therefore to be able to effectively collect, process and analyze large amounts of data in real time. To do these tasks, several existing works [3–6] use Big Data frameworks, facing the necessity of knowledge extraction. In particular, in [3] a Big Data analytics framework is presented, capable of providing a health monitoring application for an aerospace and aviation industrial. In [4] the authors use open source technologies such as Apache Spark and Kafka to implement a scalable architecture capable of processing data both online and offline. The same open source Big Data technologies are also used in [5], in order to implement an integrated Self-Tuning Engine for Predictive maintenance in Industry 4.0. The topic of predictive maintenance in a big data environment is also addressed in [6], where, with the purpose of monitoring the operation of wind turbines, a data-driven solution deployed in the cloud for predictive model generation is presented.

The applications of predictive maintenance have had a considerable diffusion with the advent of Industry 4.0, thanks to the introduction of sensors able to constantly monitor the performance of machinery. This work [7] presents how recent trends in Industry 4.0 solutions are influencing the development of manufacturing execution systems, while in [8] authors present a framework able to implement scalable, flexible and pluggable data analysis and real-time supervision systems for manufacturing environments. In [5, 9, 10] three data-driven methodologies related to the predictive maintenance services in an Industry 4.0 context are discussed.

But predictive maintenance does not only refer to failures detection, in fact it can also be used to estimate the Remaining Useful Life (RUL) of a machine. In many real-life settings, the time component (sensors measure signals that evolve over time) need to be considered, as mentioned in [11]. Possible approaches to time series data mining are wavelets, recurrent neural networks and convolutional neural networks. In [11] the authors discussed the different approaches, highlighting how neural networks work better than other models, though not significantly.

Work on a context similar to this study is described in [12], where authors present an application of an image processing system in the monitoring and control of the hot-rolling of steel bars. Unlike in that work, this study does not use visual recognition to characterise the products, but appropriate features extracted from the measurements taken by the sensors during the process.

3   Maintenance Needs and Challenges in the Steel Bar Production Industry

VDL Weweler systems are robust and deliver reliability and cost-effective operation for on-highway applications in the most demanding operating environments. Manufacturing facilities in Apeldoorn (The Netherlands) are highly automated with robotics to play an essential role in the cost-effective and reliable production.

The newly designed production line of VDL Weweler brings together a series of processes that were previously operated separately. The trailing arms are now hot-formed and then tempered in a single continuous process, in which the residual heat from the forming process can be reused for the tempering process. This process cut energy requirements by 35% and significantly reduce production time.

Another significant benefit of the new production line is the ability to perform 3D forming, making it possible to build a suspension system with fewer parts that are lighter and lend themselves better to modular construction. This whole new system integrates a series of production processes, previously performed separately due to space limitations, to form a single line.

VDL Weweler provides data for testing the data-driven solution presented in this chapter as well as the SERENA's architecture. Maintenance/repairing activities within the monitored equipment are provided as output. VDL Weweler gives technical feedback to validate the project developments and suggestions to improve the overall methodology to transfer it to other industrial sectors needing similar maintenance solutions quickly.

4   Steel Bar Production Case Study: Present and Future

The prediction of the behavior of the segments and their maintenance is of high importance. The current cycle time of trailing arms production is estimated to forty-five seconds in a working day that no unexpected failures occur. The replacement of parts takes place approximately after the presentation of 18.000 pieces, and the visual inspection of the components usually allows increasing their lifetime at least after 25.000 repeats. In this industrial scenario, the SERENA platform aims to increase the segment's lifetime, enabling predictive maintenance techniques to replace the details on the rolling machine and provide the operators with relevant information through Augmented Reality (AR) technology for maintenance operations. In parallel, by reducing the stoppages in the production line, products' quality should be improved.

The API-pro software, currently in use, manages and schedules the maintenance activities, which experienced operators usually perform. The entire replacement/maintenance process is estimated to be around one hour. Unfortunately, it is not a fixed time, as it can vary depending on the segments' temperature, and it can reach up to four hours, including the wait time for the parts to cool.

The rolling machine supplier does maintenance activities on the hydraulics unit and the device once per year. While maintenance operations are taking place, the production line stops because of serial production and the strict relationship among all the activities. The waiting time for maintenance activities on the rolling mill can vary from 1 to 4 h. Through API-Pro Software, instructions are provided by documents and/or smart devices to the maintenance personnel.

Fig. 1 VDL WEWELER production line

4.1 *Maintenance Needs*

For the SERENA pilot case, the focus is on the forming of trailing arms through a rolling mill type machine by predicting the in-time replacement of the coating segments used by the device. The trailing components are designed to suit specific vehicle models as well as detailed operational areas. Better comfort, payload, drivability, and lifetime are assured through the high quality and long service life of the VDL Weweler production line, ensuring better comfort and road-holding needed for safe and economical transportation. The trailing arms production line starts by heating steel bars of $1000 \times 100 \times 50$ (mm) in the responsible station while the rolling process follows as depicted in Fig. 1.

The current situation of the maintenance and repairing activities does not include a predictive maintenance approach. The SERENA platform discussed technically in Chapters "A hybrid Cloud-to-Edge Predictive Maintenance Platform, Data-Driven Predictive Maintenance: A Methodology Primer, and Services to Facilitate Predictive Maintenance in Industry 4.0" allows exploring this possibility in various industrial sectors. The rolling mill machine used for the forming process is monitored by applying multiple sensors to the device. The acquired data is compensated with a digital twin model or a physics-based model and eventually achieves an accurate prediction of the replacement of segments. Additionally, the main benefits focus on decreasing the downtimes, reducing the exchange of components related costs, and improving the final product quality by exploiting the segments' life-cycle efficiently.

The SERENA platform gathers data from the rolling machine within the VDL Weweler, creates models for predicting maintenance needs, correlates the rolling machine data with the digital model, and then predicts the exchange of segments. The technical feedback received from VDL Weweler during the 36-month duration improves the SERENA developments orienting the results in real industrial applications.

The most critical equipment for monitoring includes the rolling mill machine. The focus is on predicting and scheduling the replacement of the coated segments, aiming at increasing, on the one hand, their lifetime and, on the other, improving the trailing arms quality. The main three parameters that affect the segment's lifetime are high temperature, the friction between the trailing arms and components, and finally, the forces applied.

The purpose of the SERENA platform is to predict the replacement of segments of the rolling mill machine. The target is to accurately predict when the components need to be replaced through a collection of data from the milling machine's sensors and their correlation with a digital twin model. Additionally, and as the product's quality is strictly related to the milling machine's working conditions, a measuring system is designed and developed. This measuring system precisely calculates the formed trailing arms' straightness once the milling process is completed. Based on the measured values and their correlation with the milling machine's status, accurate maintenance predictions are foreseen. Moreover, the SERENA platform schedules the maintenance activities to reduce the production stoppage time and avoid any interruptions with the production plan. The maintenance operators are equipped with AR technologies for guiding them through correct task execution and training.

4.2 Equipment Description

The steel parts production pilot case is focused on the rolling milling machine in Fig. 2. This machine is composed of two rolling cylinders which are rotating through the use of torque motors. The lower rolling cylinder has a fixed position, and only the upper cylinder can move vertically. Three different geometrically coated segments are attached to the rolling cylinders. The segments are used in order to form the trailing arm by applying force. Currently, the segments are replaced after 18.000 repeats as preventative maintenance and for safety reasons in order not to completely destroy the segments.

4.3 Steel Bar Production Process

The stretching process of the rolling mill machines includes several steps. The blank entering from the heating unit is fed sequentially to the individual passes by a robot. The blank is formed analogously to the contours of the individual roller grooves. The robot arbitrarily rotates the blank by up to 180° around its longitudinal axis between the separate passes. As the rollers can be removed from the forging roll for a tool or groove change, it is possible to work with entirely circular, closed roller grooves or with grooves in the form of circular segments. The robot and the rollers work in a master-slave mode during the rolling process, with the roller rotation angle acting as the master signal. This operation mode allows shock-free working and ensures almost zero wear compared to mechanical, rigidly coupled drive systems. The robot movement is synchronized with the rollers' servo drives so that the stretched

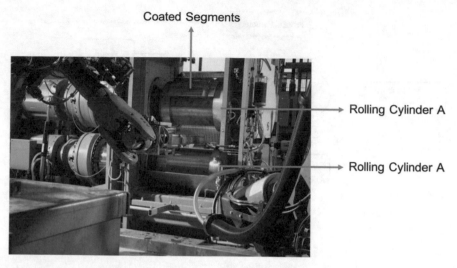

Fig. 2 Equipment description

workpiece is moved according to the prevailing peripheral speed of the rollers. Thus, during operation, the rate of workpiece movement is matched to the speed profile.

A high-level illustration of the production process is displayed in Fig. 3. As mentioned earlier, a steel metal bar is inserted in the heating station. The heating process takes approximately five minutes. Afterward, the robot R2 is responsible for picking up the heated metal bar and proceeds with the rolling mill's rolling process. For this operation, another robot, R3, is used, and the two robots are cooperating to achieve the required functions of the rolling process. At the end of the rolling process production, the robot R3 further inspects the outcome. In case of deviations in the steel bar geometry, the process parameters of the rolling mill need to be modified to reach the desirable geometrical characteristics.

5 SERENA System in the Steel Bar Production Case

5.1 Architecture

The SERENA cloud platform is built on a lightweight micro-services architecture, that allows the core cloud services and the edge gateway to be managed as a single domain. The edge gateways are located close to the factory equipment that generates the data used to perform the maintenance predictions. A schema representing the global architecture is shown in Fig. 4.

All services running on the SERENA cloud platform are implemented as Docker containers and managed by Docker Swarm. The use of containers enables modularity

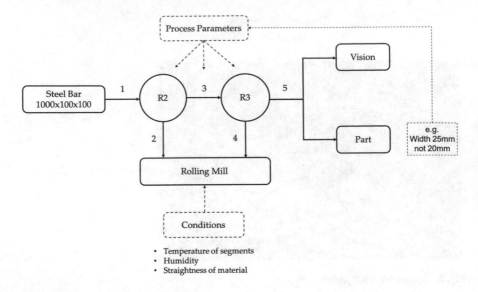

Fig. 3 Production process description

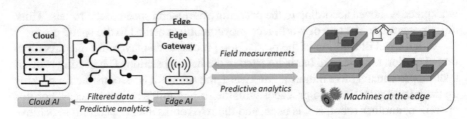

Fig. 4 High level architecture of the global system

in the SERENA cloud platform, making it simple to remove a given service and substitute it with an alternative implementation. Thus, the "plug-and-play" requirement of the SERENA platform is achieved, giving end-users the freedom to choose the technology implementation that best suits their unique needs and corporate guidelines. It also helps to future-proof the platform. Older technologies can easily be updated and/or replaced with newer alternatives as they become available without modifying the underlying platform architecture. The plug-and-play concept is also applied to the repositories, which uses a RESTapi and canonical JSON-LD message format to facilitate communication with the repositories. In cases where tight integration is required, the APIs and the data warehouses are packaged as a group of services, and the whole group can be replaced as necessary. Services communicate via HTTP RESTapi, although other protocols, like MQTT, are also supported. The RPCA (Reverse Proxy Certification Authority) middleware acts as the interface to the edge gateways and external systems, providing security and routing services. Apache NiFi serves as the central communications broker between services.

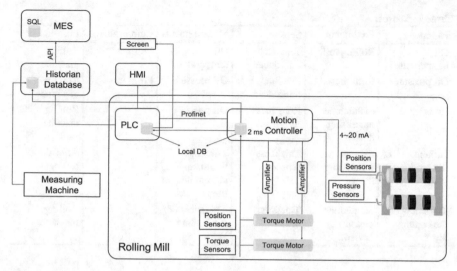

Fig. 5 Data connections

5.2 Data Connections

Figure 5 depicts the data connections for the rolling mill machine. Data from the rolling mill sensors are pre-processed to link the values collected from the force, torque, and position sensors to the rolling angle of machines' cylinders. To this aim, the data collected from the pressure, torque, and position sensors are converted into degrees in the historical data.

5.3 Data Acquisition

Several sensors have been integrated into the machine to acquire measurement over key parameters to support the inference of insight from machine processes data. Table 1 provides a list of the sensors applied in the rolling mill as well as to the standalone measuring system.

5.4 Data Analytics

The objective of the implemented methodology is to estimate the RUL (residual useful life) value of machine cylinders, as these components are expensive and they wear out differently over time, making difficult to estimate their degradation. Having a real time prediction of the machine's RUL value every time a new product is produced

Table 1 Sensors

Parameter	Equipment	Sensor	Measurements	Com protocol	Location
Oil temperature	Rolling mill	Temperature transducer	Oil temperature	Profinet	Rolling machine
Oil pressure	Rolling mill	Pressure transducer	Oil pressure	Profinet	Rolling machine
Roughness	Standalone measuring device	Roughness sensor	Oil pressure		Rolling machine
Dimensions of the product	Standalone measuring device	Thickness sensor	Thickness, width, and straightness of the product		After the rolling machine
Thickness of the coating layer	Standalone measuring device	Thickness sensor	Thickness of the coating layer		Rolling machine

is important for the company to be able to estimate the level of degradation of the machinery, thus intervening promptly with maintenance, saving time and money. The proposed methodology consists of 3 main building blocks, which are: (i) *Feature engineering*, (ii) *Data labeling*, (iii) *Model training and validation*.

5.4.1 Feature Engineering

In this block we have extracted significant features from the various input signals collected by the machinery. Each signal is summarised through the use of several features, able to characterise it. The extraction of these features has been done taking into account both the measuring machines and the rolling mill measurement.

From each input signal, we extracted the following 7 statistical features from the data collected by the measuring machine: minimum value, maximum value, mean value, standard deviation, variance, kurtosis (statistical index relating to the form of the distribution reflecting the concentration of data around its own average) and skewness (symmetry index of a distribution). In addition to this feature set we also include an eighth feature calculated as the distance between the signal collected in the measuring machine and the corresponding signal measured in the rolling mill. The error between the two series was calculated using the measure root-mean-square error (RMSE). In addition, the error of one measurement cycle also takes into account the errors of previous cycles, by adding them cumulatively, thus assuming the ever-increasing error over time.

5.4.2   Data Labeling

In some industrial context, it may be possible that no label is associated with the signals collected by the machinery. In these cases, in order to use a predictive methodology able to estimate the residual value of a machine, a previous data labelling step is required. This step is usually done manually by a domain expert, who is able through his knowledge to evaluate various signals and label them properly. However, this operation is very time-consuming and in an industry 4.0 context, where the various processes are robotized and automized, it would be useful to have an automatic methodology able to perform also the data labelling step.

So, in this section we offer two different methodologies able to automatically label the data collected by the machinery, using as a label the estimation of the degradation of the machinery itself. In particular, the label we want to assign is a decreasing numerical value that represents the RUL (residual useful life) of the machine. The smaller this value, the greater the risk of the machine breaking down. Knowing in advance the dates on which some components of the machinery have been replaced or in which maintenance has been done, we have assigned the same range of labels (from the highest to the lowest value) to each time interval between two of these consecutive dates. In this way, a cycle immediately after the replacement had a maximum label value, while a cycle just before the replacement had a minimum label value. In particular, we defined the RUL using two different strategies:

- the first strategy assigns the RUL a **linearly** decreasing trend over time;
- the second strategy assigns the RUL an **exponentially** decreasing trend over time.

The difference between these two strategies lies in the trend of the RUL over time. While the first strategy assumes that the degradation is constant during the various production cycles, the second strategy is based on the idea that in the first production cycles the performance of the machinery degrades much more slowly than in the final cycles.

In the first strategy, the formula for defining the RUL labels can be expressed as follows:

$$RUL = ceiling[(1 - X/X_{tot}) * n]$$

where X represents the current day or the current cycle (depending on whether we want to consider the RUL as linearly dependent with production cycles or with working days), X_{tot} is the total number of days or the total number of cycles, n is a parameter that we can set manually in order to define the maximum value we want to use in our labels, and ceiling is a function that can transform a floating number into its immediately greater integer.

In the second strategy, the formula defining the exponential trend of the RUL is as follows:

$$RUL = ceiling\left[-e^{\frac{ln(n+1)}{X_{tot}} * X} + (n+1)\right]$$

where the parameters take on the same meaning as in the previous formula. Here, since the generic exponential function $-e^x$ is -1 when the x value is 0, we have introduced the term $n+1$ so that when x is 0 (first production cycle considered), the RUL of the machine is assigned the maximum label value. Moreover, the fraction that multiplies X to the exponent causes the RUL to be 0 (the machinery needs an intervention) when the value on the x-axis is equal to the number of total cycles.

5.4.3   Model Training and Validation

Finally, the purpose of this block is to build a classification model capable of predicting the correct RUL value of a new cycle, based on the values learned from the historical data. The training of the model is done using the features extracted in the features computation block, in addition to the label assigned in the data labelling block (if no original labels were present). Two state-of-the-art supervised learning algorithms have been tested in these blocks: Decision Tree and Random Forest, where the most performing parameters of each algorithm are chosen thanks to a self-tuned strategy based on a grid optimisation search over the classification algorithm parameters. The proposed methodology compares the performances obtained with the two different algorithms, highlighting the best one.

To compare the performances of different models, the accuracy is given for each one of them. Moreover, for each of the predicted labels, we also report the recall and precision metrics, which are particularly useful in case the dataset is not balanced. The metrics are defined as follows:

Accuracy: Correct Predictions/Total Predictions
Recall = True Positive/(True Positive + False Positive)
Precision = True Positive/(True Positive + False Negative)

5.5   Results

This section shows the results obtained on the analyzed use case using the techniques described in Sect. 5.4. Here, the purpose is to verify how well the real-time prediction methodology performs on the data under analysis, using both the approaches reported in 5.4.2 to label the data.

The available data are composed of 9756 production cycles collected between 22/01/2020 and 24/02/2020. During this period, the segment was replaced on the machine at two different times: 04/02/2020 and 19/02/2020. These maintenance dates allow us to divide the production cycles into three different production groups within which the RUL of the machinery will assume all the values from the maximum (first cycle) to the minimum (cycle close to maintenance). In order to correctly assign the RUL labels, we need to know for each group the start date and end date of production. Because when we collected the data, the third production group was not yet finished, we have considered in our training model only the first two groups: a total of 5233 production cycles collected between 22/01/2020 and 19/02/2020.

Since the number of cycles is not very high and we have only two complete production groups available, we used the leave one out cross validation (loocv) technique to test the performance of the predictive model. Loocv allows us to use a single production cycle as a test set, and all other cycles as a training set. This operation is repeated iteratively, using each time a different production cycle as a test set. Then, at each iteration a new predictive model is built to predict the RUL value of the cycle used in the test set. So, at the end of this process, each cycle was used once in the test set and was labeled with a predictive RUL value.

In the following paragraphs, for each proposed configuration, the results are compared with those obtained from a baseline methodology, a more traditional approach to address feature engineering in use cases similar to the one discussed here. This baseline methodology adopts the same approach as ours with regard to data labelling; instead, in the feature engineering block, the only features extracted are the seven statistics: cumulative error between rolling mill measurements is not considered.

All the following experiments were conducted using the current production cycle as the X value in the formula introduced in Sect. 5.4.2. This means that we consider the life of the machinery to be decreasing over the production cycles.

5.5.1 Experiment with Linear Approach

Since in this case we assume that the RUL has a linear trend over time, each production group will be divided into n subgroups of equal size (production cycles), where n is the number of labels we want to use in our experiments. We have manually set this parameter to 10 and 100. Tables 2 and 3 show the accuracy results obtained using these two configurations, while the confusion matrices shown in Figs. 6 and 7 refer to the case with 10 labels using the Decision Tree classifier. The columns represent the predicted labels, while the rows represent the real labels. Values in bold on the diagonal are the correct predictions.

From these results we can see that only the statistical features (baseline) or only the cumulative error are not sufficient input variables to guarantee a good performance of the predictive model, while their combination allows to reach very good results. The low level of accuracy obtained by using only the error as an input variable can be explained by the fact that very different signals can be equally distanced from

Table 2 Accuracy using 10 labels

	Decision tree	Random forest
cumulative error (c.e)	0.26	0.30
Baseline	0.68	0.72
Baseline + c.e.	0.99	0.97

Table 3 Accuracy using 100 labels

	Decision tree	Random forest
Cumulative error (c.e.)	0.19	0.22
Baseline	0.28	0.32
Baseline + c.e.	0.93	0.69

	1	2	3	4	5	6	7	8	9	10
1	**425**	37	10	30	12	8	0	0	0	0
2	34	**388**	61	20	20	0	0	0	1	0
3	15	60	**338**	70	35	1	0	3	1	0
4	32	19	82	**293**	76	5	1	13	2	0
5	7	17	42	82	**334**	27	0	14	0	1
6	5	0	2	4	29	**402**	78	0	1	2
7	0	0	1	2	0	90	**325**	53	41	11
8	0	0	2	15	12	3	70	**322**	68	31
9	0	0	0	1	1	2	32	57	**357**	74
10	0	0	0	0	4	1	17	37	76	**389**

| RECALL | 0.81 | 0.74 | 0.64 | 0.56 | 0.63 | 0.76 | 0.62 | 0.61 | 0.68 | 0.74 |
| PRECISION | 0.82 | 0.74 | 0.62 | 0.56 | 0.63 | 0.74 | 0.62 | 0.64 | 0.65 | 0.76 |

Fig. 6 Baseline confusion matrix

a Rolling Mill's signal measurements; so the error alone is not able to effectively characterize the signal. The same thing can be said if we use the baseline approach with only the statistical features, as signals with different trends can be characterized by very similar set of features.

The best performance obtained using both variables is confirmed in Fig. 7: from this confusion matrix we can see that only 47 cycles out of a total of 5233 were not correctly labeled. In addition, most of the errors are near the diagonal of the matrix, indicating that almost all of the errors made by the classifier occurred between adjacent classes. A cycle with a high residual value has never been labeled with a low label, or vice versa.

Using a higher number of labels, obviously the possibilities of error increase and consequently the accuracy values decrease, but, as shown in Table 3, the results are still good and in line with what was obtained with 10 labels.

	1	2	3	4	5	6	7	8	9	10
1	521	1	0	0	0	0	0	0	0	0
2	1	521	1	0	0	1	0	0	0	0
3	0	2	520	1	0	0	0	0	0	0
4	0	0	1	521	1	0	0	0	0	0
5	0	0	0	2	519	3	0	0	0	0
6	0	0	0	0	2	519	2	0	0	0
7	0	0	0	2	0	3	514	2	2	0
8	0	0	0	0	0	1	4	516	2	0
9	0	0	0	0	0	1	3	2	515	3
10	0	0	0	0	0	0	0	0	4	520
RECALL	0.99	0.99	0.99	0.99	0.99	0.99	0.98	0.98	0.98	0.99
PRECISION	0.99	0.99	0.99	0.99	0.99	0.98	0.98	0.99	0.98	0.99

Fig. 7 New methodology confusion matrix

Table 4 Labels percentage distribution

Label	10	9	8	7	6	5	4	3	2	1
Occurrences(%)	28.9	16.8	12.0	9.3	7.6	6.5	5.6	4.9	4.5	3.9

5.5.2 Experiment with Exponential Approach

The purpose of this section is to evaluate the performance of the predictive model, assuming that the degradation of the machinery has an exponentially decreasing trend. Labeling the production cycles based on this hypothesis, the dataset will no longer be balanced; in fact, since the degradation will increase much faster at the end of the production group, the cycles that will be assigned a high RUL value will be more numerous than those with a low RUL value. In particular, each class will contain always fewer elements than the previous class. Table 4 contains the percentages of elements that have been assigned to each class after the data labelling step.

Also in this case the results obtained are very good and are on average even better than those obtained in the linear case. Table 5 contains the accuracy values with the different variables used as inputs and we can see that with our methodology the performance of the predictive model is improved compared with other conditions, as also demonstrated by the two confusion matrices shown in Figs. 8 and 9.

A comparison between the matrices shows that the baseline methodology performs well when it has to recognise the optimal condition of the machinery (label 9 or 10), as, due to the exponential trend in degradation, these labels are the most numerous. However, when degradation increases and the labels defining the RUL of the machinery rapidly change, this methodology is no longer able to assign each

Table 5 Accuracy value using exponential approach

	Decision tree	Random forest
Cumulative error (c.e.)	0.39	0.42
Baseline	0.70	0.75
Baseline + c.e.	0.99	0.97

	1	2	3	4	5	6	7	8	9	10
1	**122**	30	28	8	1	2	4	12	0	0
2	32	**144**	20	1	8	5	15	5	0	0
3	19	17	**141**	53	6	6	11	4	0	0
4	9	2	45	**167**	36	11	17	5	0	0
5	3	8	0	36	**194**	51	32	12	0	0
6	1	2	7	17	60	**203**	63	40	1	4
7	1	11	11	10	25	78	**247**	81	0	22
8	13	4	4	6	16	35	90	**412**	32	17
9	0	0	0	0	0	2	1	40	**718**	123
10	0	0	0	0	1	5	19	20	133	**1336**

| RECALL | 0.59 | 0.62 | 0.54 | 0.57 | 0.57 | 0.51 | 0.50 | 0.65 | 0.81 | 0.88 |
| PRECISION | 0.61 | 0.66 | 0.55 | 0.56 | 0.55 | 0.51 | 0.49 | 0.65 | 0.81 | 0.88 |

Fig. 8 Baseline confusion matrix

production run to the correct label, as demonstrated by the recall and accuracy values in Fig. 8. Instead, using our new approach, the predictive model is able to correctly recognise the labels not only of the initial classes, but also of those in which degradation is greater, like shown in Fig. 9.

5.6 System Deployment

The proposed data-driven methodology has been implemented in a python package and integrated as a service in the SERENA cloud platform as described in Chapter "A hybrid Cloud-to-Edge Predictive Maintenance Platform" and Chapter "Services to Facilitate Predictive Maintenance in Industry 4.0". The cloud platform is built upon lightweight micro-services bridging the gap between the edge and the cloud. The services enabling its functionalities are implemented as Docker containers enabling scalability and modularity. Whilst edge devices usually consist of a separate network

	1	2	3	4	5	6	7	8	9	10
1	207	0	0	0	0	0	0	0	0	0
2	1	227	1	1	0	0	0	0	0	0
3	0	1	254	2	0	0	0	0	0	0
4	0	0	2	289	1	0	0	0	0	0
5	0	0	0	1	335	0	0	0	0	0
6	0	0	0	0	1	396	1	0	0	0
7	0	0	0	2	0	1	484	1	0	0
8	0	0	0	1	0	0	2	624	2	0
9	0	0	0	0	0	2	0	3	878	1
10	0	0	0	0	0	0	0	0	4	1510
RECALL	1	0.98	0.98	0.98	0.99	0.99	0.99	0.99	0.99	0.99
PRECISION	0.99	0.99	0.98	0.98	0.99	0.99	0.99	0.99	0.99	0.99

Fig. 9 New methodology confusion matrix

from the cloud infrastructure, cloud services and edge gateways along with edge devices/sensors can be managed as a single domain.

Edge components are deployed at the edge gateway as docker containers. All the services are centrally managed via the Docker Swarm. The use of containers allows for a single service to be easy to replace in terms of functionality and/or technology. Therefore, the cloud platform with its functionalities is implemented onto a stack of indicative technologies that can be replaced with alternative ones, granting its user with the freedom to select the technologies that suit his/her specific needs.

The communication throughout the system is enabled using HTTP REST APIs and canonical JSON-LD messages without excluding other types of communication such as MQTT. The JSON data format is partially built upon the MIMOSA open standard for asset maintenance and, in particular, the CRIS 3.2 schema. The JSON messages are forwarded from the edge gateways to the cloud via RPCA, creating a security middleware between edge gateways and external systems. A central communication broker is implemented using Apache NiFi.

The platform is designed and implemented to store and process vast volumes of semi-structured data. To this end, the Apache Hadoop framework has been adopted, allowing for distributed processing across computer clusters. Data are consumed by integrated applications for analytic, scheduling, visualization purposes that can be extended to include additional functionalities. In this work, the focus is placed on the analytics service.

6 Discussion and Conclusion

The steel products manufacturing industry represents one of the SERENA use cases where the platform's innovative solutions were deployed and tested. The main cloud platform was instantiated to best address the specific requirements of the use case, while the focus was paid upon data driven analytics.

The aforementioned use case provides a testbed for testing and validating the proposed analytics methodology towards detecting abnormal events and imminent failures. It should be noted that while the SERENA solutions were fitted in the existing production facilities relatively seamlessly, the proper training of the AI algorithms was challenging. However, the results obtained on a collection of real-data are promising. There is a clear improvement over the performance obtained with the traditional approach. This indicates that considering both rolling mill and measuring machine measurements can bring significant benefit to the predictive model. Moreover, the proposed methodology is general purpose and takes into account different degradation functions over time. A methodology like the one proposed in this chapter is able to adapt to the RUL prediction of different components with different degradation trends.

The proposed approach can therefore be decoupled from the specific use-case on which it was tested in this chapter, generalising it to other contexts of industrial steel production.

Acknowledgements This research has been partially funded by the European project "SERENA – VerSatilE plug-and-play platform enabling REmote predictive mainteNAnce" (Grant Agreement: 767561).

References

1. T. Cerquitelli, D.J. Pagliari, A. Calimera, L. Bottaccioli, E. Patti, A. Acquaviva, M. Poncino, Manufacturing as a data-driven practice: Methodologies, technologies, and tools, in *Proceedings of the IEEE* (2021)
2. J. Lee, H.D. Ardakani, S. Yang, B. Bagheri, Industrial big data analytics and cyber-physical systems for future maintenance & service innovation. Proc. CIRP **38**, 3 (2015)
3. B. Xu, S.A. Kumar, Big data analytics framework for system health monitoring, in *2015 IEEE International Congress on Big Data* (2015), pp. 401–408. https://doi.org/10.1109/BigDataCongress.2015.66
4. G.M. D'silva, A. Khan, Gaurav, S. Bari, Real-time processing of iot events with historic data using apache kafka and apache spark with dashing framework, in *2017 2nd IEEE International Conference on Recent Trends in Electronics, Information Communication Technology (RTEICT)* (2017), pp. 1804–1809. https://doi.org/10.1109/RTEICT.2017.8256910
5. D. Apiletti, C. Barberis, T. Cerquitelli, A. Macii, E. Macii, M. Poncino, F. Ventura, istep, an integrated self-tuning engine for predictive maintenance in industry 4.0, in *IEEE International Conference on Parallel & Distributed Processing with Applications, Ubiquitous Computing & Communications, Big Data & Cloud Computing, Social Computing & Networking, Sustainable Computing & Communications, ISPA/IUCC/BDCloud/SocialCom/SustainCom 2018,*

Melbourne, Australia, December 11-13, 2018, ed. by J. Chen, L.T. Yang (IEEE, 2018), pp. 924–931

6. M. Canizo, E. Onieva, A. Conde, S. Charramendieta, S. Trujillo, Real-time predictive maintenance for wind turbines using big data frameworks, in *2017 IEEE International Conference on Prognostics and Health Management (ICPHM)* (2017), pp. 70–77. https://doi.org/10.1109/ICPHM.2017.7998308

7. S. Jaskó, A. Skrop,T. Holczinger, T. Chován, J. Abonyi, Development ofmanufacturing execution systems in accordance with industry 4.0requirements: a review of standard-and ontology-based methodologiesand tools. Comput. Ind. **123**, 103300 (2020). https://doi.org/10.1016/j.compind.2020.103300. http://www.sciencedirect.com/science/article/pii/S0166361520305340

8. R.S. Peres, A. Dionisio Rocha, P. Leitao, J. Barata, Idarts—towards intelligent data analysis and real-time supervision for industry 4.0, Computers in Industry **101**, 138 (2018). https://doi.org/10.1016/j.compind.2018.07.004. http://www.sciencedirect.com/science/article/pii/S0166361517306759

9. S. Proto, F. Ventura, D. Apiletti, T. Cerquitelli, E. Baralis, E. Macii, A. Macii, Premises, a scalable data-driven service to predict alarms in slowly-degrading multi-cycle industrial processes, in *2019 IEEE International Congress on Big Data, BigData Congress 2019, Milan, Italy, July 8-13, 2019*, ed. by E. Bertino, C.K. Chang, P. Chen, E. Damiani, M. Goul, K. Oyama (IEEE, 2019), pp. 139–143

10. F. Ventura, S. Proto, D. Apiletti, T. Cerquitelli, S. Panicucci, E. Baralis, E. Macii, A. Macii, A new unsupervised predictive-model self-assessment approach that scales, in *2019 IEEE International Congress on Big Data, BigData Congress 2019, Milan, Italy, July 8-13, 2019*, ed. by E. Bertino, C.K. Chang, P. Chen, E. Damiani, M. Goul, K. Oyama (IEEE, 2019), pp. 144–148

11. S. Munikoti, L. Das, B. Natarajan, B. Srinivasan, Data-driven approaches for diagnosis of incipient faults in dc motors. IEEE Trans. Ind. Inf. **15**(9), 5299 (2019). https://doi.org/10.1109/TII.2019.2895132

12. P. O'Leary, Machine vision for feedback control in a steel rolling mill. Comput. Ind. **56**(8), 997 (2005). https://doi.org/10.1016/j.compind.2005.05.023. http://www.sciencedirect.com/science/article/pii/S0166361505001351. Machine Vision Special Issue

In-Situ Monitoring of Additive Manufacturing

Davide Cannizzaro, Antonio Giuseppe Varrella, Stefano Paradiso,
Roberta Sampieri, Enrico Macii, Massimo Poncino, Edoardo Patti,
and Santa Di Cataldo

Abstract Additive Manufacturing, in great part due to its huge advantages in terms
of design flexibility and parts customization, can be of major importance in maintenance engineering and it is considered one of the key enablers of Industry 4.0.
Nonetheless, major improvements are needed towards having additive manufacturing solutions achieve the quality and repeatability standards required by mass production. In-situ monitoring systems can be extremely beneficial in this regard, as
they allow to detect faulty parts at a very early stage and reduce the need for post-
process analysis. After providing an overview of Additive Manufacturing and of the
state of the art and challenges of in-situ defects monitoring, this chapter describes
an in-house developed system for detecting powder bed defects. For that purpose, a
low-cost camera has been mounted off-axis on top of the machine under considera-

D. Cannizzaro (✉) · A. G. Varrella · M. Poncino · E. Patti · S. Di Cataldo
Department of Control and Computer Engineering, Politecnico di Torino, Turin, Italy
e-mail: davide.cannizzaro@polito.it

A. G. Varrella
e-mail: antoniogiuseppe.varrella@studenti.polito.it

M. Poncino
e-mail: massimo.poncino@polito.it

E. Patti
e-mail: edoardo.patti@polito.it

S. Di Cataldo
e-mail: santa.dicataldo@polito.it

E. Macii
Interuniversity Department of Regional and Urban Studies and Planning, Politecnico di Torino,
Turin, Italy
e-mail: enrico.macii@polito.it

S. Paradiso · R. Sampieri
Stellantis, Turin, Italy
e-mail: stefano.paradiso@stellantis.com

R. Sampieri
e-mail: roberta.sampieri@stellantis.com

© Springer Nature Singapore Pte Ltd. 2021
T. Cerquitelli et al. (eds.), *Predictive Maintenance in Smart Factories*,
Information Fusion and Data Science,
https://doi.org/10.1007/978-981-16-2940-2_10

tion. Moreover, a set of fully automated algorithms for computer vision and machine learning enables allow the timely detection of a number of powder bed defects along with the layer-by-layer monitoring of the part's profile.

1 Introduction

Additive manufacturing (AM) is the process of joining materials layer by layer to create objects, starting from a three-dimensional (3D) model. AM uses computer-aided-design (CAD) software to drive a 3D printer towards creating precise geometric shapes. Afterwards, the CAD model is converted into a series of layers and instructions suitable for a printer, and then the layers are printed sequentially. AM is one of the most promising manufacturing technologies [1] that are emerging in the context of Industry 4.0. The first machine for AM was developed in 1984 by Chuck Hull [2], the founder of 3D Systems, a well-known company that develops 3D printers. The first prototype was expensive thus making it inaccessible to the wide market. However, the constant improvements in hardware and software solutions, affecting their performance and reducing their cost, has results in the spread of 3D printers in many industrial fields [3]: in the past few years, many industries have already embraced AM technologies, and they are beginning to incorporate AM in their production lines and business models [4]. This is especially true for the medical, automotive and aerospace sectors, where AM is pushing forward innovative applications such as custom-made implants and prosthetic, as well as lightweight and complex components of cars and airplanes [5–8].

The main advantages of AM over traditional technologies are mass customization, on-demand and decentralized manufacturing, freedom of design, and the ability to manufacture complex structures and fast prototyping. In other words, AM technology allows mass customization at low cost, implying that industries can design and personalize parts production with small efforts and without lengthy delivery time, or even have parts printed directly at the local distributors or service providers' premises, reducing the delivery time and the logistics requirements [9].

As it makes it possible to produce critical spare parts in small quantities at a very low cost and even to print such parts locally, AM technologies, together with predictive analytics and digitalization, could revolutionize the business models of maintenance industry and fulfill the potentials of condition-based maintenance (CbM) and predictive maintenance (PdM): in this scenario, heterogeneous sensors can be used to continuously monitor the condition of critical parts, and data-driven predictive models can be exploited to anticipate part failures and breakdowns, as well as to plan local repairs or replacements of the damaged parts by means of AM, with tremendous reduction of costs and time.

On the one hand, AM is considered one of the pillars of smart manufacturing and maintenance industry. On the other hand, it is still in the early stage of development [10]. Hence, it has some important drawbacks: build size limitations, long production time, high equipment and maintenance costs, along with a lack of quality assurance and process repeatability. Typically, an industrial 3D printer has a build volume ranging from 100 to 400 mm^3, which only allows printing objects with a small size and packed effectively [11]. Due to this limitation, some parts require to be manufactured in segments and assembled at the end. Additionally, the average time to produce a single part is high [12], which currently makes AM more suitable to mass customization manufacturing than mass production. Moreover, 3D printer associated costs are higher when compared to traditional subtractive equipment. In this context, research efforts in the field of predictive maintenance of AM machines, which are currently at the very early stages, are expected to gain momentum in the future.

As mentioned before, one of the major barrier in the widespread of AM is the lack of quality assurance and repeatability. In particular, due to the complexity of additive processes and of the parts to be produced, some defects might occur during the printing process. To guarantee the quality standards required by industries (especially in critical sectors like aerospace, automotive and healthcare) expensive and difficult post-process inspection are necessary. In order to address the aforementioned in an effective way, monitoring and control systems can be adopted for supervising the AM process during the layering procedure, enabled by sensors integrated with to the 3D printing machine. The real-time analysis of those sensor data can be useful for the early detection of defective parts, or even to correct the process in order to proactively address defects from occurring. In this regard, Machine Learning and Computer Vision (CV) analytics can significantly contribute.

Empowered by the aforementioned, the main focus of this chapter is in-situ defects monitoring in one of the most diffused metal AM technology, that is Direct Metal Laser Sintering (DMLS). In the following sections, a brief introduction of the main AM technologies available in the market is provided (Sect. 2), along with the state of the art in the field of monitoring systems (Sect. 3). Then, follows a near real-time AM monitoring system applied in an industrial use case, using visible-range cameras and computer vision algorithms (Sect. 4). Finally, the experimental results are discussed (Sect. 5) to conclude with Sect. 6.

2 Additive Manufacturing: Process and Methods

The main steps of a typical AM process are illustrated below in Fig. 1. It starts from a 3D CAD model of the products, then the model is converted into a stereolythography (STL) file format that describes the surface geometry of the object to be printed. The STL is processed by a slicer, a software that converts the model into a series of thin layers, and produces instructions tailored to a specific AM system. Finally, the final product may undergo a subtractive finishing process to achieve higher quality.

In AM, many methods and materials are used to meet the demand for printing complex structures at fine resolutions. Figure 2 shows the main technologies and the respective materials used. Stereolithography, Multi Jet Fusion, and Selective Laser Melting are intended for plastic materials. Instead, Fused Deposition Modelling and

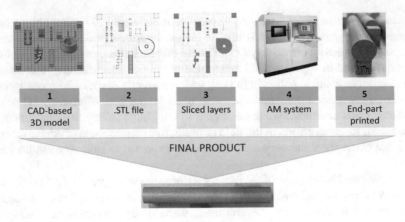

Fig. 1 AM process: main phases

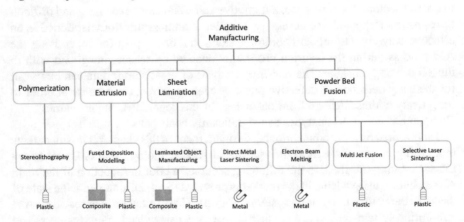

Fig. 2 AM technologies

Laminated Object Manufacturing use both composite and plastic materials, while DMLS and Electron Beam Melting are suitable only for metal products.

The *Polymerization* process uses ultraviolet light to transform a plastic polymer from liquid to solid.

- *Stereolithography* is a liquid-based process that uses ultraviolet light (or electron beam) to initiate a chain reaction on a layer of photosensitive polymer [13]. Typically, a post-process treatment such as heating or photo-curing is required for some printed parts in order to achieve the desired mechanical performance. Stereolithography is suitable for high-quality parts with a fine resolution as low as 10 μm [14]. However, it is relatively slow and expensive, suitable only for limited materials like resins that change their structure after intense exposure to ultraviolet light.

Material Extrusion is a process where a spool of material goes through a heated nozzle in a continuous stream to build a 3D object.

- *Fused Deposition Modeling* uses a continuous filament of a thermoplastic polymer to print an object. The process is based on the extrusion of a heated thermoplastic filament through a nozzle tip that deposits precisely the material on the platform to build the part. The main advantages are that no chemical post-processing or curing is required leading to an overall reduction of the costs [15]. However, the process does not allow high resolutions (<0.25 mm) with respect to other AM processes.

Sheet lamination is an AM methodology where thin sheets of material are bonded together to form a single piece that is cut into the desired 3D object.

- *Laminated Object Manufacturing* is one of the first commercial methods ever developed. It is based on a layer-by-layer cutting and lamination of sheets or rolls of material [16]. This process allows to reduce the cost of tooling and manufacturing time, and it is one of the most used AM methods for larger structures. However, Laminated Object Manufacturing has some disadvantages like inferior surface quality (without post-processing) and lower dimensional accuracy compared to the powder bed methods.

Powder Bed Fusion (PBF) involves spreading powder material on top of the previous layers through a recoater, with a reservoir providing new material supply. A heat source, typically a laser or electron beam, selectively melts together each layer of powder [17].

- *Direct Metal Laser Sintering* (DMLS) is one of the most used PBF technologies in AM, and hence is the main focus of this chapter. It allows to print parts with a 95% density without requiring any additional post-build sintering [18]. The DMLS process was developed by EOS GmbH, and it was first introduced in the EOS M250 machine in 1995. It uses a laser that is directly exposed to the metal powder in liquid phase sintering. Figure 3 depicts the schematic picture of the instrument used for DMLS. The chamber is filled with an inert gas to avoid oxidation of the

Fig. 3 DMLS set-up

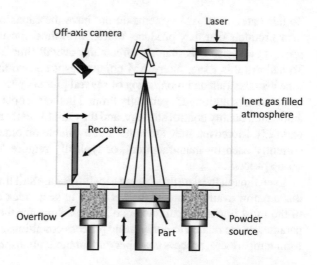

powder. The powder bed is heated to almost the melting point of the material and it is controlled by a piston that is lowered the same amount of the layer thickness each time a layer is finished. The powder allocated in the powder source chamber is spread using a recoater. The excess powder is removed and then the laser fused the powder at a specific location for each layer specified by the design. Nowadays, DMLS is the most widespread technique for metal, thanks to its trade-off between printing accuracy and cost [19].

- *Electron Beam Melting* is a relatively novel process that involves an electron laser beam powered by high voltage in the range of 30–60 kV to melt the powder [20]. The process is similar to DMLS but, in this process, the part is printed in a high vacuum chamber to avoid oxidation. Due to the high power produced by the laser, the penetration depth in the powder bed is greater with respect to DMLS. This high power could lead to the formation of cracks on the surface of the material, reducing the process stability.
- *Multi Jet Fusion* is an AM method developed by Hewlett-Packard. It creates parts additively thanks to a multi-agent printing process. Multi Jet Fusion manufacturing technique is particularly useful to create unique plastic parts with a good surface finish [21]. However, only a few plastic materials are suitable for the Multi Jet Fusion method.
- *Selective Laser Sintering* (SLS) is another printing process similar to DMLS, where a powder is sintered or fused by applying a laser beam. While DMLS is suitable only for metals, SLS can be used with various polymers. The main disadvantage compared to DMLS is that the accuracy is limited by the size of particles of the material. However, it allows a fine resolution and high quality for printing complex structures [22].

3 In-Situ Monitoring: Current Solutions

To this date, many AM systems do not have the capability to assess the quality of their products that they produce, unless with time-consuming and expensive post-process analysis. This majorly affects the overall time and cost of the production. To address this issue, many AM companies are providing software for near real-time visualization and monitoring of several process parameters. Nonetheless, these commercial solutions are generally limited in their scope: they do not develop a fully automated quality control strategy, and they fail to detect minor defects in the printed part [23]. Moreover, they are either not available on older machine models that are currently used by manufacturers, or typically require an expensive and complex set-up process.

Even though the data processing capabilities are still limited, most of the machines that are now available on the market allow to keep track of the history and behavior of the most important parameters of the machine for the entire job process. These parameters are related to either basic process conditions like laser power, build platform temperature, process chamber and ambient atmosphere, recoater speed, etc. or machine conditions like cooling system status, electrical power levels, powder source level, etc. [24]. Up-to-date machines can also be equipped with increasingly complex hardware like high-speed and infrared cameras, thermocouples, pyrometers, photo-detectors, that allow to monitor directly or indirectly many process parameters that are involved into the generation of defects.

Taking advantage of such advanced sensors, researchers are using an increasing amount of data analytics approaches to address the problem of AM monitoring and control, mostly combining off-line Machine Learning (ML) methods with other types of algorithms, depending on the nature of the sensor data [25–27].

With the use of optical sensors (either in the visible or in the Infra-Red range) to acquire images of the build chamber, many approaches combine ML together with Computer Vision algorithms to perform AM defect analysis. For example, in [28] the authors propose a Convolutional Neural Network (CNN) for autonomous detection and classification of different type of anomalies. They extract the relevant features from the images and train the algorithm to detect possible part damages, incomplete spreading of powder, and recoating errors. In [29], the authors propose an online monitoring algorithm that uses computer vision to detect defect formation in every layer of the PBF process. They collect a set of images with a high-resolution camera, and analyze the images to detect layers with low quality of fusion or defects. In [30], the authors utilized a high-speed camera to acquire the powder bed images. Then, they combined a Support Vector Machines with a CNN to analyze the images and detect possible defects during the printing process.

Other researches focus on more unconventional types of data sources. For example, in [31, 32] PBF defects are detected based on acoustic emissions acquired by either microphones of optical fibers, using different types of neural networks like Deep Belief Networks or Spectral Convolutional Neural Networks for the classification task. Finally, in [33], heterogeneous data from a combination of different sensors are collected from the laser system, powder bed and recoating process in an Electron Beam Melting system, and then fed into a Support vector data description (SVDD) model to find major deviations from the expected pattern (e.g., cracks and holes on the surface). A disadvantage of the latter approaches is that they generally require invasive modification on the machine and expensive hardware additions for the data collection.

Recent literature demonstrates data-driven techniques being promising to monitor and improve AM processes. Nonetheless, the application to real-world industrial scenarios is often limited by a number of factors. First, by the need of complex and expensive hardware. Second, by current limitations in the efficient collection, storage, annotation, and integration of the sensed data. Moreover, ML techniques, and especially deep learning, typically require to be trained on a large amount of annotated data (e.g., for image-based systems, a large amount of images of the powder bed per each possible defect and condition). Considering that the intentional production of defective parts is not viable in industry, as it is costly and time-consuming, the generation of a significant training set becomes an issue.

To address this problem, ML can be put at work even to create synthetic and realistic images, resembling the ones obtained during the print of a defective part [34]. In this regard, one of the most promising approaches is Generative Adversarial Network (GAN), a generative model characterized by training a pair of neural networks (a generator and a discriminator) in competition with each other [35]. Among the many families of GANs recently proposed by literature, Conditional GAN (CGAN) uses an additional input in the generator network to direct the generation process [36], obtaining very good results in many data augmentation tasks. While GANs are becoming very popular in many Computer Vision tasks, the application to the AM sector is still limited. In [37], the authors developed a CGAN to produce synthetic data using as input the layer-by-layer images captured with a near-infrared high-resolution camera. As conditional images, the authors use a synthetic picture of the shape of the built artifact, and then they generate the corresponding near-infrared image, which mimics the characteristics of the captured ones. Ideally, a similar approach may be used for more complex tasks involving a large number of defects, like the one addressed by our case-study. More specifically, starting from an original image without flaws, CGAN can generate synthetic images of defects, augmenting the training set for defect classification. Unfortunately, the amount of data needed to train a standard CGAN for this purpose would not be much lesser than the one needed to train a classifier from scratch. To overcome this problem, in the present study a very promising architecture called ConSinGAN [38] is employed, that is able to learn generative models from a single training image (more details will follow in the next sections).

At present, the use of real-time monitoring systems in AM, coupled with Machine Learning models and, eventually, data augmentation strategies, is still mainly applied to the early detection of layer defects, for part qualification purposes. Given the lack of maturity of AM for mass production, little attention has been devoted so far to monitoring the status and health of the AM machine, for predictive maintenance purposes. Nonetheless, the extensive undergoing research in the field of sensing technology and ML models for AM defect analysis and process characterization is creating a solid background for future developments, where similar concepts may be easily applied to the early detection of machine anomalies and failure prediction. This is expected to enabled significant advantages in terms of reduced unnecessary post-processing analysis, equipment replacement, increased process safety, availability, and efficiency.

4   Case-Study: Layer-Wise Defects Monitoring System for DMLS Based on Visible-Range Imaging

In this section, a fully-automated framework for layer-wise defects monitoring in DMLS AM is presented. The system is based upon an off-axis low-cost optical sensor for image acquisition of the powder bed and the manufactured object during the layering process. Then, a set of fully-automated tools based on Computer Vision and ML allow the detection of possible defects of the part that are hard to spot by visual inspection. The prototype was designed in a real-world industrial scenario and

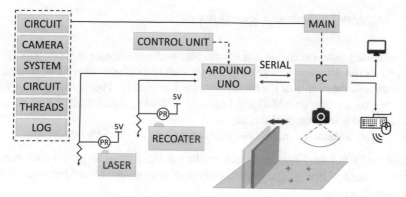

Fig. 4 In-situ monitoring system

developed on top of a fully operative DMLS machine, in an automotive company. By allowing the early stopping/correction of the faulty artefacts, this system is expected to improve process repeatability and majorly reduce human intervention, with major positive impacts on the production costs.

Together with the data collection and the automated defects detection methods, a first prototype of synthetic image generation leveraging ConSinGAN is proposed, which starts from the images acquired with the proposed fully-automated framework to produce synthetic images addressing two different types of powder bed defects.

4.1 Data Acquisition and Processing System

The implemented monitoring system is built upon low-cost hardware and camera on top of an EOS M290 DMLS printer in an automotive company. As shown in Fig. 4, it includes:

- an Arduino Uno computing platform directly connected with the 3D printer used to manage the system, trigger the camera, and take images of the powder bed.
- an IDS UI-1540-SE 1.31Mpix camera (1280 × 1024 resolution). The camera is triggered through the Application Programming Interface (API) made available by the manufacturer. The acquisition is off-axis with respect to the optical path of the laser.
- A standard PC running Linux, to collect images in the Portable Network Graphics (PNG) format with a resolution of 1280 × 1024 and run the image analytics algorithms.

The image acquisition is automatically triggered by 3D printer states, exploiting the signals emitted respectively by the action of the laser and of the recoater, by means of photo-resistors. By doing so, the system is able to acquire images of the powder bed before and after each layer is printed, without requiring any user interaction.

4.2 Defects Detection Algorithms

The proposed solution includes a set of near real-time image analytic algorithms that allows the detection of five different powder bed defects, as well as continuous monitoring of the profile of the object that is being printed. The algorithms are based on image processing and Machine Learning and were developed in Python using OpenCV and Keras standard libraries.

Figure 5a–e show five main categories of defects targeted by our system:

- *Holes* (Fig. 5a): localised lacks of metallic powder that create small dark areas in the powder bed image. The origin is a lack of powder due to bad regulation of the dosing factor.
- *Spattering* (Fig. 5b): droplets of melted metal ejected from the melt pool and landed in the surroundings.
- *Incandescence* (Fig. 5c): high-intensity areas in the completed layer image, resulting from an excess of laser energy density and a consequent inability by the melt pool to cool down correctly.
- *Horizontal* defects (Fig. 5d): dark horizontal lines in the powder bed caused by the incorrect spreading of the powder, possibly because of the geometric imperfection of the piece or of the metallic powder.
- *Vertical* defects (Fig. 5e): vertical undulation of the powder bed, consisting of alternated dark and light lines along the direction of the recoater's path. The origin is either a mechanical interference between the part and the recoater or a mechanical defect of the recoater's surface.

Each of these powder bed defects is known to cause either porosities or microstructural alterations in the printed parts, as well as lower mechanical characteristics.

The pipeline for run-time defects detection consists of several image processing steps.

- *Normalization* (Fig. 6a): the images are first normalised against a common reference frame, in order to correct uneven illumination problems. To do so, an image of the powder bed is acquired before the start of the layering process and used as a reference throughout.
- *Contrast enhancement* (Fig. 6b): a standard background subtraction algorithm is applied to make the objects more distinguishable from each other, as well as from the background [39].
- *Objects identification* (Fig. 6c): intensity discontinuities are identified by means of automated intensity thresholding algorithm. This provides a rough identification of the different objects in an image.
- *Morphological filtering* (Fig. 6d): specific objects are recognized based on their shape, exploiting morphological algorithms. More in detail, Watersheds, and Hough transform, followed by standard morphological regularization (i.e., opening, closing, holes filling), are respectively applied to identify round-shaped and horizontal/vertical lines. Based on the specific shape and number of detected objects, the software identifies a specific category of defect.

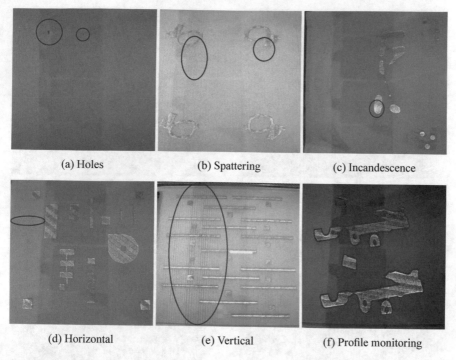

(a) Holes   (b) Spattering   (c) Incandescence

(d) Horizontal   (e) Vertical   (f) Profile monitoring

Fig. 5 **a–e** Examples of powder bed defects targeted by the system. **f** Profile monitoring example

As an example, in Fig. 6 is presented the pipeline applied to detect a spattering defect, showing the outcome of each intermediate step. Spattering is indeed one of the defects that most frequently happen during powder bed fusion: it involves tiny particles of liquid metal being ejected from the laser's path, which may contaminate the powder bed and create issues such as porosity, roughness, and lack of adhesion in the finished parts. At the end of the last step in Fig. 6 (i.e. morphological filtering), the most relevant spatters are identified.

4.3 Part Profile Monitoring Algorithm

Besides powder bed defects detection, the system includes a fully automated *profile monitoring* suite that is able to monitor the profile of the build on a layer-by-layer basis (see an example in Fig. 5f). This task has additional algorithmic and computational challenges compared to basic powder bed defects because the printed parts may have very different shapes and dimensions.

In our solution, profile monitoring is addressed as a semantic segmentation problem. Semantic segmentation aims to cluster parts of an image together, which belong to the same object, using a pixel-level prediction to classify each pixel in an image

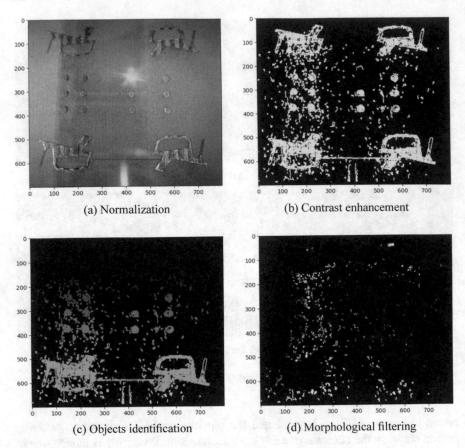

(a) Normalization   (b) Contrast enhancement

(c) Objects identification   (d) Morphological filtering

Fig. 6 Example of spattering defects detection pipeline

according to a category. In other words, image segmentation becomes a binary classification task, where each pixel needs to be labeled as belonging to the object of interest (in our case, the printed part) or the powder bed.

To achieve this purpose, a U-Net architecture is employed [40], a deep learning algorithm that was initially designed for biomedical image segmentation and then successfully applied to many different Computer Vision applications. The network implements an end-to-end fully convolutional network (FCN) composed of convolutional and pooling layers without any dense layer, which makes it suitable for any image size. As shown in Fig. 7, the architecture is composed of two paths. The first path (encoding) is the contraction path or encoder used to capture the context in the image. It consists of a stack of various convolutional and max-pooling layers with gradually decreasing feature map dimension. The second path (decoding) is a symmetric expanding path or decoder used to enable precise localization using transposed convolutions. In the presented approach, the U-Net follows the implementation suggested by Ronneberger et al. [40]. The encoding path consists of the

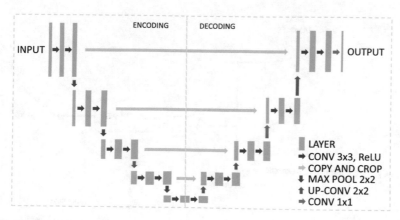

Fig. 7 U-Net architecture

repeated application of two 3×3 convolutions, with Rectified Linear Unit (ReLU) as activation function and a 2×2 max pooling operation. The decoding path consists of an upsampling of the feature map (up-conv) followed by a 2×2 convolution with ReLU as activation function, a concatenation with the correspondingly cropped feature map from the contraction path, and two 3×3 convolutions, each followed by a ReLU. The copy and crop are necessary due to the loss of border pixels in every convolution. At the final layer, a 1×1 convolution is used to map each feature vector to the desired number of classes. In total, the network has 23 convolutional layers.

In this application, the two classes to be classified are the foreground (i.e., the printed part) and the background (i.e., the powder bed). First, it was used as input images with a size of 360×480. Then, the input size was fine-tuned on a validation set consisting of images of a representative PBF layer, and finally set the optimal size to 320×320.

4.4 Dataset Augmentation with GAN

Unfortunately, training ML algorithms to AM defect classification requires a large number of training images, with and without defects. As anticipated in Sect. 3, this is difficult to obtain in a real-world industrial setting, especially for the defect categories. To overcome this issue and increase the number of defective images available for the training, the discussed solution adopts an extended version of Conditional Generative Adversarial Network, ConSinGAN [38]. While traditional GANs still require large datasets for the training, ConSinGAN is able to generate synthetic data starting from a single image. This characteristic is of major importance, given the very limited number of defective images.

As shown in Fig. 8, a classic GAN architecture is composed of two Neural Networks, the Generator and the Discriminator, that compete to produce a generative

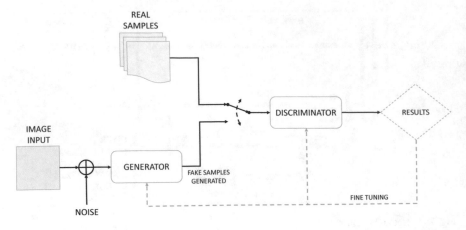

Fig. 8 Generative adversarial network architecture

model. The generator network produces synthetic data starting from the input, with the aim of obtaining data that resembles the real and available samples. At the same time, the discriminator network is a binary classifier that is trained to distinguish the synthetic samples produced by the generator from the real ones. The training phase consists of an adversarial process between the two networks, in which the generator tries to increase the error rate of the discriminator, while the latter tries to distinguish real and synthetic data. Hence, the core idea is to implement an indirect training of the generative model through the discriminator: the generator has no direct access to real data, and interacts only with the discriminator [41].

The ConSinGAN is structured in a multi-stage and multi-resolution approach, as shown in Fig. 9. First the ConSinGAN uses, as input for few iterations, an image with a coarse resolution to learn mapping a random noise vector z to a low-resolution image (See *Generator Phase 0* in Fig. 9). After the initial training has converged, the size of the generator is increased, adding three additional convolutional layers. Each stage takes as input the raw features from the previous stage, and a residual connection is used in each stage to feedforward the last convolutional layer (See *Stage 1* in Fig. 9). The process is repeated N times until the desired output resolution is reached. The synthetic image created by the generator is given as input to the discriminator, together with a real image from the dataset. The discriminator is trained to discern if the images given are real or synthetic, comparing them.

The most critical part of the synthetic image generation is the image harmonization (see an example in Fig. 10). It consists in transforming a composite image, called naive (Fig. 10b), into a realistic image by applying to the changes made the same style and appearance of the original image, called training (Fig. 10a). The output of this process is a synthetic image similar to the original one (Fig. 10c). To perform this process, the network is first trained to learn a generative model from the original image. Then, given the naive image as input, it tries to transform it into an image that should resemble the original learned distribution. Once the model is trained on a

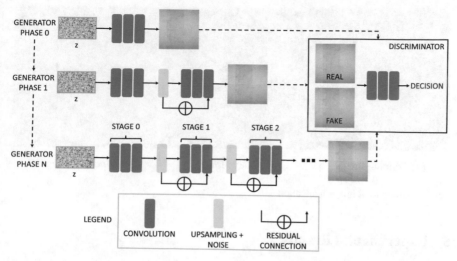

Fig. 9 ConSinGAN architecture [38]

(a) Training image   (b) Naive image   (c) Harmonized image

Fig. 10 Example of image harmonization [38]

specific image, various synthetic images can be generated by varying only the naive picture.

Fig. 11 shows an example of AM image generation with the ConSinGAN. To generate new images presenting defects, the image harmonization process is carried out as follows: (i) the generator is trained using a real image captured without defects (Fig. 11a); (ii) a naive image is created applying modifications to the training image by using a photo editing software (Fig. 11b); (iii) the naive images is harmonized to resemble the training image characteristics (Fig. 11c). In this way, the synthetic defects are assimilated into the original image.

The data generated can be used to increase the dataset available to both the train and validation phases of defect detection algorithms, breaking down the costs of generating a huge set of real defect data.

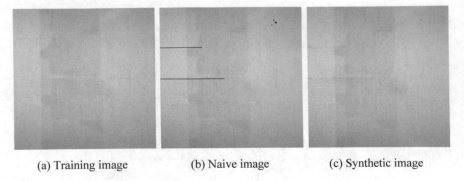

(a) Training image   (b) Naive image   (c) Synthetic image

Fig. 11 Example of image generation

5 Experimental Results

To validate the defect detection algorithms, a set of pre-annotated images was used with all the defects targeted by the proposed system. For the five main categories of defects, we run a statistical validation by analyzing whether the algorithm identified the defect or not, using metrics that are widely accepted in classification tasks:

- Accuracy: it represents the number of correct classifications with respect to the total cases.

$$Accuracy = \frac{TP + TN}{TP + TN + FP + FN}$$

- Precision: it is the fraction of relevant cases among the retrieved instances.

$$Precision = \frac{TP}{TP + FP}$$

- Recall: it is the fraction of the total amount of relevant instances that were retrieved.

$$Recall = \frac{TP}{TP + FN}$$

In this work, True Positives (TP) represent the instances when the algorithms were able to detect a defect that was really present. True Negatives (TN) represent the instances when a given defect was not present, and the algorithm was right in not detecting it. False Positive (FP) and False Negative (FN) represent the possible errors of the algorithms, respectively in detecting a defect that was not present, or not being able to identify a defect that was present. Table 1 reports the results obtained on a test set of 24 images with different powder bed conditions and defects.

For all the five defects, the results of the metrics considered are ≥75% with the worst results for the Incandescence defects (Precision: 75%) and Vertical defects (Recall: 75%). According to our tests, Incandescence proved to be the most chal-

Table 1 Defects detection algorithms validation

	Holes	Spatt.	Incand.	Horizontal	Vertical
TP	14	22	15	11	6
TN	8	1	4	11	16
FP	2	1	5	1	0
FN	0	0	0	1	2
Accuracy (%)	91.3	95.8	79.2	91.6	91.67
Precision (%)	87.5	95.6	75	91.6	100
Recall (%)	100	100	100	91.6	75

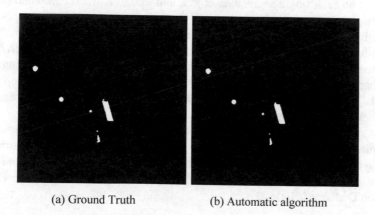

(a) Ground Truth   (b) Automatic algorithm

Fig. 12 Profile monitoring: validation example

lenging defect to be recognized, probably due to the high pixel luminosity variation. On the other hand, Spattering defects are the easiest due to the high amount of spatters generated.

For the profile monitoring task, the validation exploits the Sørensen–Dice coefficient (DSC) to compare the profile segmentation obtained with our algorithm against a manually obtained ground truth. This metric is used to gauge a 0 to 1 similarity of two binary images, as follows:

$$DSC = \frac{2|X \cap Y|}{|X| + |Y|}$$

where $|X|$ and $|Y|$ are the number of pixels of the two images (in our case, the automatic segmentation and the ground truth) and $|X \cap Y|$ the number of pixels that are common to both images. Figure 12 shows an example of this procedure, with (a) the binary mask obtained by manual segmentation, used as the ground truth, and (b) the binary mask obtained by our profile monitoring suite.

Table 2 Mean execution time of the algorithms

Operation	Time (s)	Operation	Time (s)
Holes	0.791	Horizontal	0.593
Spattering	0.574	Vertical	0.932
Incandescence	0.821	Profile monitoring	2.461

In the conducted tests, a very good similarity between automated segmentation and manual ground truth was obtained, with mean DSC value equal to 0.878, when computed on each single segmented object, and to 0.911, when computed on each layer image taken as a whole.

Finally, Table 2 reports the execution time of all the tested algorithms. As can be seen from the reported values, execution times are all below 2.5 s, which is well below the time elapsing between two subsequent layers. Profile monitoring is the algorithm taking the longest time (2.461 s) because it involves running a deep neural network. The other algorithms, which exploit standard image processing operations, are all below 1 s of execution.

As final step, the results of the synthetic data generation process were validated. In the implemented prototype, the image augmentation was preliminarily carried out on two types of defects: holes and horizontal. We assessed the results obtained by the ConSinGan through a most widely used GAN evaluation metric, that is the Fréchet Inception Distance (FID) [42]:

$$FID = ||\mu_X - \mu_Y||^2 + Tr(\Sigma_X + \Sigma_Y - 2\sqrt{\Sigma_X \Sigma_Y})$$

where X is the set of real images, Y is the set of synthetic images; μ_X and μ_Y are the feature-wise mean of the real and generated images, respectively; Σ_X and Σ_Y are the covariance matrix for the real and generated feature vectors, respectively; Tr refers to the trace operation in matrix, which is the sum of the diagonal elements; $||\mu_X - \mu_Y||^2$ refers to sum squared difference between the two mean vectors.

The FID value ranges between 0 and plus infinite: a value near to 0 means that the two images are the same, while a very high value means that the two images are completely different.

Figure 13 shows an example of a real image without defects (a) and a synthetic image (b) with holes defects, as generated by our approach. As it can be seen, the two images are quite similar, corresponding a FID value of 85.74.

Table 3 reports the results obtained after 2000 steps, on a test set of 20 generated images with holes and horizontal defects. As expected, the FID shows similar results for all the generated images. This is due to the fact that the synthetic images were generated starting from same training set, and that the added defects are quite similar. The FID values of the synthetic images are also quite small, demonstrating a good similarity to real ones with the same defects.

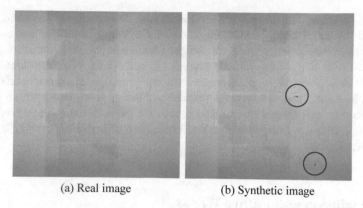

(a) Real image (b) Synthetic image

Fig. 13 Comparison between a training image and a synthetic one with defects (holes)

Table 3 ConSinGAN results

Synthetic image	1	2	3	4	5	6	7	8	9	10
FID	131.07	156.45	93.66	85.74	97.74	107.27	103.71	119.70	93.49	99.81
Synthetic image	11	12	13	14	15	16	17	18	19	20
FID	110.82	90.14	102.64	117.35	127.00	101.58	112.89	98.61	111.51	103.21

Table 4 Defects detection results on artificially generated images

	Holes	Horizontal
TP	8	8
TN	14	12
FP	1	2
FN	1	2
Accuracy (%)	91.7	83.3
Precision (%)	88.9	80.0
Recall (%)	88.9	80.0

As a final experiment, the synthetic images were tested with the defects detection algorithms. Table 4 reports the results obtained on a test set of 24 synthetic images with holes and horizontal defects, generated with the ConSinGAN.

For both the defects, the results of the considered metrics (Accuracy, Precision, and Recall) are ≥80% with the worst results for the Horizontal defects (Precision and Recall: 80%). Hence, according to the experiments conducted, the defect detection results are similar to the results obtained with real images. The holes detection algorithm has better Accuracy and Precision with synthetic images (Accuracy: 91.6%, Precision: 88.9%) than with real images (Accuracy: 91.3%, Precision: 87.5%), while the Recall is better with the real images (Recall with synthetic images: 88.9%, Recall with real images: 100%). Instead, the horizontal detection algorithm has accuracy, precision and recall better with real images (Accuracy: 91.6%, Precision: 91.6%, Recall: 91.6%) than with synthetic images (Accuracy: 83.3%, Precision: 80.0%, Recall: 80.0%) Sure indeed, holes are easier to detect compared to horizontal defects, that can be easily confused with the powder bed background. However, on the test case with synthetic images, the metrics results for the horizontal defect were similar to those with real images, with a small reduction of performance ≤12% for the worst-case Recall.

6   Conclusion and Future Works

Computer Vision and ML have proven to be promising approaches for the in-situ monitoring of the AM process. Nonetheless, the actual use of these approaches in a real industrial scenario is still limited due to a number of challenges. First of all, the lack of effective data collection infrastructure specifically devoted to AM. Second, the necessity to train the models on large annotated datasets, which are typically costly and difficult to obtain in industrial environments.

This chapter presented a low-cost camera-based in-situ defects monitoring system for metal PBF and data generation. The preliminary prototype, developed and tested in a real industrial scenario of an automotive company, includes: (i) a set of run-time Computer Vision and ML algorithms to detect five different categories of powder bed defects, as well as the layer-wise monitoring of the profile of a printed part and (ii) a synthetic image generation model, for data augmentation purposes. Experimental results suggest that the algorithms have a good performance in terms of defect detection accuracy and profile segmentation and they are suitable for near real-time execution with low-cost hardware. The framework is currently being extended to provide layer-by-layer comparisons between the profile of the printed part (as returned by the profile monitoring suite) and the desired profile as defined by the slicer. This will allow a near real-time automated detection of any profile alterations during a build.

In its current form, the presented framework only allows the continuous monitoring of the AM part, and the timely detection of macroscopic defects based on the real-time comparison with a baseline. Nonetheless, the continuous growth of collected images is creating a solid ground for future extensions in the direction of predictive analytics. In this regard, it is planned to integrate the available image dataset with process parameters as well as with post-process information (i.e., results of quality and mechanical tests on the finished parts, as well as historical information about part breakdowns), in order to train models able to predict and possibly correct the failure of a part, applying process optimization strategies.

References

1. European Commission. Additive manufacturing in fp7 and horizon 2020 (2014)
2. H. Lipson, M. Kurman, *Fabricated: The New World of 3D Printing* (Wiley & Sons, 2013)
3. D.S. Thomas, S.W. Gilbert, Costs and cost effectiveness of additive manufacturing. NIST Spec. Publ. **1176**, 12 (2014)
4. T. Pereira, J.V. Kennedy, J. Potgieter, A comparison of traditional manufacturing vs additive manufacturing, the best method for the job. Proc. Manuf. **30**, 11–18 (2019)
5. M. Delic, D.R. Eyers, The effect of additive manufacturing adoption on supply chain flexibility and performance: An empirical analysis from the automotive industry. Int. J. Prod. Econom. **228**, (2020)
6. L.J. Kumar, C.G. Krishnadas Nair, Current trends of additive manufacturing in the aerospace industry, in *Advances in 3D Printing & Additive Manufacturing Technologies* (Springer, 2017), pp. 39–54
7. M. Salmi, K.-S. Paloheimo, J. Tuomi, J. Wolff, A. Mäkitie, Accuracy of medical models made by additive manufacturing (rapid manufacturing). J. Cranio-Maxillofacial Surgery **41**(7), 603–609 (2013)
8. S.-S. Yoo, 3d-printed biological organs: medical potential and patenting opportunity (2015)
9. J.V.L. Silva, R.A. Rezende, Additive manufacturing and its future impact in logistics. IFAC Proc. **46**(24), 277–282 (2013)
10. M. Attaran, The rise of 3-d printing: the advantages of additive manufacturing over traditional manufacturing. Bus. Horizons **60**(5), 677–688 (2017)
11. L.J.P. de Araújo, E. Özcan, J.A.D. Atkin, M. Baumers, C. Tuck, R. Hague, Toward better build volume packing in additive manufacturing: classification of existing problems and benchmarks, in *Annual International Solid Freeform Fabrication Symposium* (Austin, Texas, USA, 2015)
12. A. Hussein, L. Hao, C. Yan, R. Everson, P. Young, Advanced lattice support structures for metal additive manufacturing. J. Mater. Proc. Technol. **213**(7), 1019–1026 (2013)
13. P.F. Jacobs, *Rapid prototyping and manufacturing: fundamentals of stereolithography* (Soc, Manuf Eng, 1992)
14. X. Wang, M. Jiang, Z. Zhou, J. Gou, D. Hui, 3d printing of polymer matrix composites: a review and prospective. Compos. Part B: Eng. **110**, 442–458 (2017)
15. C.K. Chua, K.F. Leong, C.S. Lim. *Rapid Prototyping: Principles and Applications (with Companion CD-ROM)*. World Scientific Publishing Company (2010)
16. B. Mueller, D. Kochan, Laminated object manufacturing for rapid tooling and patternmaking in foundry industry. Comput. Ind. **39**(1), 47–53 (1999)
17. N. Guo, M.C. Leu, Additive manufacturing: technology, applications and research needs. Front. Mech. Eng. **8**(3), 215–243 (2013)
18. M.W. Khaing, J.Y.H. Fuh, L. Lu, Direct metal laser sintering for rapid tooling: processing and characterisation of eos parts. J. Mater. Proc. Technol. **113**(1–3), 269–272 (2001)
19. I. Campbell, O. Diegel, J. Kowen, T. Wohlers. Wohlers report 2018: 3D printing and additive manufacturing state of the industry: annual worldwide progress report. Wohlers Associates (2018)
20. L.E. Murr, S.M. Gaytan, D.A. Ramirez, E. Martinez, J. Hernandez, K.N. Amato, P.W. Shindo, F.R. Medina, R.B. Wicker, Metal fabrication by additive manufacturing using laser and electron beam melting technologies. J. Mater. Sci. Technol. **28**(1), 1–14 (2012)
21. F.N. Habib, P. Iovenitti, S.H. Masood, M. Nikzad, Fabrication of polymeric lattice structures for optimum energy absorption using multi jet fusion technology. Mater. Des. **155**, 86–98 (2018)
22. Z. Agarwala, D. Bourell, J. Beaman, H. Marcus, J. Barlow, Direct selective laser sintering of metals. Rapid Prototyp. J. (1995)
23. V. Carl, Monitoring system for the quality assessment in additive manufacturing, in *AIP Conference Proceedings*, vol. 1650 (American Institute of Physics, 2015), pp. 171–176
24. M. Mani, B.M. Lane, M Alkan Donmez, S.C. Feng, S.P. Moylan, A review on measurement science needs for real-time control of additive manufacturing metal powder bed fusion processes. Int. J. Prod. Res. **55**(5), 1400–1418 (2017)

25. I. Baturynska, O. Semeniuta, K. Martinsen, Optimization of process parameters for powder bed fusion additive manufacturing by combination of machine learning and finite element method: a conceptual framework. Procedia CIRP **67**, 227–232 (2018)
26. Z. Li, Z. Zhang, J. Shi, W. Dazhong, Prediction of surface roughness in extrusion-based additive manufacturing with machine learning. Robot. Comp.-Integ. Manuf. **57**, 488–495 (2019)
27. Z. Zhu, N. Anwer, Q. Huang, L. Mathieu, Machine learning in tolerancing for additive manufacturing. CIRP Annals **67**(1), 157–160 (2018)
28. L. Scime, J. Beuth, A multi-scale convolutional neural network for autonomous anomaly detection and classification in a laser powder bed fusion additive manufacturing process. Add. Manuf. **24**, 273–286 (2018)
29. M. Aminzadeh, T.R. Kurfess, Online quality inspection using Bayesian classification in powder-bed additive manufacturing from high-resolution visual camera images. J. Intell. Manuf. **30**(6), 2505–2523 (2019)
30. Y. Zhang, G.S. Hong, D. Ye, K. Zhu, J.Y.H. Fuh, Extraction and evaluation of melt pool, plume and spatter information for powder-bed fusion am process monitoring. Mater. Des. **156**, 458–469 (2018)
31. S.A. Shevchik, C. Kenel, C. Leinenbach, K. Wasmer, Acoustic emission for in situ quality monitoring in additive manufacturing using spectral convolutional neural networks. Add. Manufact. **21**, 598–604 (2018)
32. D. Ye, G.S. Hong, Y. Zhang, K. Zhu, J.Y.H. Fuh, Defect detection in selective laser melting technology by acoustic signals with deep belief networks. Int. J. Adv. Manuf. Technol. **96**(5–8), 2791–2801 (2018)
33. M. Grasso, F. Gallina, B.M. Colosimo, Data fusion methods for statistical process monitoring and quality characterization in metal additive manufacturing. Procedia CIRP **75**, 103–107 (2018)
34. K. Alexopoulos, N. Nikolakis, G. Chryssolouris, Digital twin-driven supervised machine learning for the development of artificial intelligence applications in manufacturing. Int. J. Comput. Integrated Manuf. **33**(5), 429–439 (2020)
35. I. Goodfellow, J. Pouget-Abadie, M. Mirza, B. Xu, D. Warde-Farley, S. Ozair, A. Courville, Y. Bengio, Generative adversarial nets. Advances in Neural Information Processing Systems 2672–2680 (2014)
36. M. Mirza, S. Osindero, Conditional generative adversarial nets. arXiv preprint arXiv:1411.1784 (2014)
37. C. Gobert, E. Arrieta, A. Belmontes, B.R. Wicker, F. Medina, B. McWilliams, Conditional generative adversarial networks for in-situ layerwise additive manufacturing data, in *Proceeding of the 29th international Solid Freeform Fabrication Symposium* (2019)
38. T. Hinz, M. Fisher, O. Wang, S. Wermter, Improved techniques for training single-image gans. arXiv preprint arXiv:2003.11512 (2020)
39. W.A. Mustafa, H. Yazid, Image enhancement technique on contrast variation: a comprehensive review. J. Telecommun. Electron. Comput. Eng. (JTEC) **9**(3), 199–204 (2017)
40. O. Ronneberger, P. Fischer, T. Brox, U-net: convolutional networks for biomedical image segmentation, in *International Conference on Medical image computing and computer-assisted intervention* (Springer, 2015), pp. 234–241
41. M. Arjovsky, L. Bottou, Towards principled methods for training generative adversarial networks. *arXiv preprint* arXiv:1701.04862
42. M. Heusel, H. Ramsauer, T. Unterthiner, B. Nessler, S. Hochreiter, Gans trained by a two time-scale update rule converge to a local nash equilibrium. Adv. Neural Inf. Proc. Syst. **30**, 6626–6637 (2017)

Glossary

Accuracy   Metric for evaluating classification models. It represents the number of correct predictions, divided by the total number of predictions.

As-a-service Model   Describes a subscription based offering of latest IT computing sources meeting the needs of business customers while avoiding the need for expensive infrastructure and/or implementations.

Belt Tensioning   The level of tensioning between the motor and the gearbox of the RobotBox.

Big Data   The term is used to describe large amounts of data that cannot be efficiently stored or processed using traditional data management tools.

Certification Authority   A Certification Authority is an entity that issues digital certificates. A digital certificate certifies the ownership of a public key. This allows others to rely upon signatures made by the private key that corresponds to the certified public key. A Certification Authority acts as a trusted third party.

Class Imbalance   Refers to the hidden and latent relationship among data from a huge set of samples.

Cloud Computing   A technical solution and system architecture providing on-demand computing resources and data storage from remote servers, without an active intervention by the user.

Condition-Based Maintenance   Maintenance approach in which interventions are scheduled and updated based on the estimated current machine or production asset's condition. The machine state can be evaluated at scheduled intervals, on request, or continuously.

Corrective Maintenance   Maintenance strategy in which corrective action is taken only when a fault or failure occurs. It can be immediate or deferred.

Cyber-Physical System (CPS)   The term refers a highly interconnected system of computation, communication, control and physical elements.

Data-Driven Analytics   Set of techniques which employs statistical and machine learning techniques to extract implicit knowledge available from past measurements. They can be broadly classified in the two categories of descriptive and predictive analytics.

© Springer Nature Singapore Pte Ltd. 2021
T. Cerquitelli et al. (eds.), *Predictive Maintenance in Smart Factories*,
Information Fusion and Data Science,
https://doi.org/10.1007/978-981-16-2940-2

Downtime   Period of time when a machine/system is not available.

Edge Analytics Application   Application in which at least a part of the data processing and analysis is performed close to where the data is collected.

Failure Mode & effects (or & criticality) analysis   They are both methodologies to identify potential failure modes for a product or process, assess the risk associated to those failure modes, rank the identified issues in terms of importance and identify corrective actions to address the most serious of them.

Fault Prediction/Detection   Set of techniques that estimates the probability that a given asset will fail within a predetermined time horizon (prediction) or is in a current state of failure (detection)

Foaming Machine   Machine providing the Polyurethane Foaming. The operation performed by this machine is the core step in the manufacturing of refrigerators.

HDFS   Stands for the Apache Hadoop Distributed File System, scalable and portable, written in java, and capable of handling large datasets running on commodity hardware.

Industry 4.0   The term denotes the fourth industrial revolution, under which physical and digital technologies are combined to transform the way companies manufacture, improve and distribute their products.

Information and Communication Technologies   The term refers to a set of technologies that provide access to information through telecommunication. It is similar to IT (Information Technology), but with a focus on communication technologies.

Injection Pattern   It is A well defined time-based strategy to add, at every timestamp, a precise amount of data characterized in a specific way, for example with a label.

Internet of Things (IoT)   The term describes a network of physical objects capable of communicating and exchanging data over the internet.

Kalman Filter   It is an algorithmic linear quadratic estimation to some unknown variables given a time-series of measurements.

JavaScript Object Notation for Linked Data (JSON-LD)   The term refers a World Wide Web Consortium Recommendation to encode linked data using JSON, designed to require as little effort as possible by developers to transform their existing JSON to JSON-LD. JSON-LD allows data to be serialized similarly to traditional JSON.

Machine Learning   A subset of Artificial Intelligence (AI) that instead of following static rules coded in a program, it identifies patterns in its inputs evolving allow its algorithms to adapt over time.

MIMOSA   Non-profit organisation developing supplier-neutral standards for asset lifecycle management. MIMOSA standards support functional and interoperability requirements for managing critical infrastructure.

Model Drift   Phenomenon by which the performance of a machine learning model is degraded, due to the fact that the statistical properties of the streaming input data deviate from the original training set.

Original Equipment Manufacturer   It denotes a company whose products are marketed by its customers.

Plug-and-play   It is used to denote a device or computer bus that facilitates the discovery and use of a newly connected hardware component in a system without the need for physical device configuration.

Precision   Metric for evaluating classification models. It is is defined as the fraction of relevant instances among all retrieved instances. It can assume values from 0 to 1, where 1 is the best value.

Predictive Maintenance   Maintenance strategy based on monitoring the condition of the machinery to identify in advance various signs of deterioration, anomalies and equipment performance problems.

Programmable Logic Controller (PLC)   The term refers a programmable microprocessor-based small computer that supports data communication from other devices, allows the processing of the data and triggers outputs based on pre-programmed parameters that define its decision making and control logic that is used for automating industrial processes.

Proxy server   A proxy server is a computer server or router or software system running on a computer that acts as a gateway between an end-point client device and another server from which the client is requesting a service.

Recall   Metric for evaluating classification models. It is the fraction of retrieved instances among all relevant instances. It can assume values from 0 to 1, where 1 is the best value.

Remaining Useful Life (RUL)   Time left to the end of a machine or production asset's operational life, with respect to the functions for which it was constructed or purchased for.

RESTful   It is an application programming interface that is based upon the REpresentational State Transfer (REST) architecture which uses a subset of HTTP. It is commonly used in the creation of web services.

Reverse Proxy   A reverse proxy is a type of proxy server that is accessible from the public internet, it often keeps a cache of static content and it can provide compression or TLS encryption to the clients.

Root Mean Square Error (RMSE)   It is a frequently used measure of the differences between values. It represents the square root of the mean of the square of all of the error.

RobotBox   The COMAU's testbed consisting of a motor, a gearbox reducer, a transmission belt between the motor and the gearbox reducer, an encoder mounted on the motor to read its real time position, a 5 kilograms weight used to stress more the engine, a slicer to change the distance between the motor and the gearbox reducer, which simulates the changing the belt tensioning, and a millesimal dial gauge.

Rolling Milling Machine   Machine composed of two rolling cylinders which are rotating through the use of torque motors. It is used for the forming process in steel bar production.

Self-labelling   An unsupervised analytics methodology that adds labels on data that natively do not have.

Service Orchestration   Service orchestration is an architecture that serves to coordinate and manage computer systems, including their automated configuration, possibly spanning multiple cloud vendors, and domains, with the goal of aligning the business requests with the applications, data, and infrastructure, to meet application performance goals using minimized costs and maximized performance within budget constraints.

Small and Medium-Sized Enterprise (SME)   Enterprises whose personnel number is below a certain limit. According to the EU recommendation 2003/361, the main factors determining if an enterprise is an SME is personnel number and either turnover of balance sheet total.

Virtual Private Network (VPN)   The term describes a method by which two endpoints (networks or device and network) communicate in a secure, encrypted way through a public network.

Wasserstein Distance   Proximity function used for comparing the probability distributions of two variables.

Wiener-Process   Refers to a real-valued continuous-time stochastic process. It can also be defined as a Gaussian process with zero expectation and co-variance function. A standard Wiener process is also called Brownian motion, since it gave a representation of a Brownian path.

Index

© Springer Nature Singapore Pte Ltd. 2021
T. Cerquitelli et al. (eds.), *Predictive Maintenance in Smart Factories*,
Information Fusion and Data Science,
https://doi.org/10.1007/978-981-16-2940-2

Printed in the United States
by Baker & Taylor Publisher Services